Birkhäuser

Lecture Notes in Applied and Numerical Harmonic Analysis

More information about this series at http://www.springer.com/series/4968

David R. Adams

Morrey Spaces

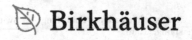

David R. Adams
Department of Mathematics
University of Kentucky
Lexington, KY, USA

ISSN 2296-5009   ISSN 2296-5017   (electronic)
Lecture Notes in Applied and Numerical Harmonic Analysis
ISBN 978-3-319-26679-4   ISBN 978-3-319-26681-7   (eBook)
DOI 10.1007/978-3-319-26681-7

Library of Congress Control Number: 2015954629

Mathematics Subject Classification (2010): 31A15, 31B15, 43A15, 46E35

Springer Cham Heidelberg New York Dordrecht London

Printed on acid-free paper

Springer International Publishing AG Switzerland is part of Springer Science+Business Media (www.
springer.com)

Dedicated to my wife and daughter,
  Jeanie Artis Adams
  and
  Sissy B. Meredith, Ph.D.

LN-ANHA Series Preface

The *Lecture Notes in Applied and Numerical Harmonic Analysis (LN-ANHA)* book series is a subseries of the widely known *Applied and Numerical Harmonic Analysis (ANHA)* series. The Lecture Notes series publishes paperback volumes, ranging from 80-200 pages in harmonic analysis as well as in engineering and scientific subjects having a significant harmonic analysis component. *LN-ANHA* provides a means of distributing brief-yet-rigorous works on similar subjects as the *ANHA* series in a timely fashion, reflecting the most current research in this rapidly evolving field.

The *Applied and Numerical Harmonic Analysis (ANHA)* book series aims to provide the engineering, mathematical, and scientific communities with significant developments in harmonic analysis, ranging from abstract harmonic analysis to basic applications. The title of the series reflects the importance of applications and numerical implementation, but richness and relevance of applications and implementation depend fundamentally on the structure and depth of theoretical underpinnings. Thus, from our point of view, the interleaving of theory and applications and their creative symbiotic evolution is axiomatic.

Harmonic analysis is a wellspring of ideas and applicability that has flourished, developed, and deepened over time within many disciplines and by means of creative crossfertilization with diverse areas. The intricate and fundamental relationship between harmonic analysis and fields such as signal processing, partial differential equations (PDEs), and image processing is reflected in our state-of-the-art *ANHA* series.

Our vision of modern harmonic analysis includes mathematical areas such as wavelet theory, Banach algebras, classical Fourier analysis, time-frequency analysis, and fractal geometry, as well as the diverse topics that impinge on them.

For example, wavelet theory can be considered an appropriate tool to deal with some basic problems in digital signal processing, speech and image processing, geophysics, pattern recognition, bio-medical engineering, and turbulence. These areas implement the latest technology from sampling methods on surfaces to fast algorithms and computer vision methods. The underlying mathematics of wavelet theory depends not only on classical Fourier analysis, but also on ideas from abstract harmonic analysis, including von Neumann algebras and the affine group. This leads to a study of the Heisenberg group and its relationship to Gabor systems, and of the metaplectic group for a meaningful interaction of signal decomposition methods.

The unifying influence of wavelet theory in the aforementioned topics illustrates the justification for providing a means for centralizing and disseminating information from the broader, but still focused, area of harmonic analysis. This will be a key role of *ANHA*. We intend to publish with the scope and interaction that such a host of issues demands.

Along with our commitment to publish mathematically significant works at the frontiers of harmonic analysis, we have a comparably strong commitment to publish major advances in applicable topics such as the following, where harmonic analysis plays a substantial role:

Bio-mathematics, Bio-engineering,   *Image processing and super-resolution;*
   and Bio-medical signal processing;   *Machine learning;*
Communications and RADAR;   *Phaseless reconstruction;*
Compressive sensing (sampling)   *Quantum informatics;*
   and sparse representations;   *Remote sensing;*
Data science, Data mining,   *Sampling theory;*
   and Dimension reduction;   *Spectral estimation;*
Fast algorithms;   *Time-frequency and Time-scale analysis*
Frame theory and noise reduction;   *– Gabor theory and Wavelet theory.*

The above point of view for the *ANHA* book series is inspired by the history of Fourier analysis itself, whose tentacles reach into so many fields.

In the last two centuries Fourier analysis has had a major impact on the development of mathematics, on the understanding of many engineering and scientific phenomena, and on the solution of some of the most important problems in mathematics and the sciences. Historically, Fourier series were developed in the analysis of some of the classical PDEs of mathematical physics; these series were used to solve such equations. In order to understand Fourier series and the kinds of solutions they could represent, some of the most basic notions of analysis were defined, e.g., the concept of "function." Since the coefficients of Fourier series are integrals, it is no surprise that Riemann integrals were conceived to deal with uniqueness properties of trigonometric series. Cantor's set theory was also developed because of such uniqueness questions.

A basic problem in Fourier analysis is to show how complicated phenomena, such as sound waves, can be described in terms of elementary harmonics. There are two aspects of this problem: first, to find, or even define properly, the harmonics or spectrum of a given phenomenon, e.g., the spectroscopy problem in optics; second, to determine which phenomena can be constructed from given classes of harmonics, as done, for example, by the mechanical synthesizers in tidal analysis.

Fourier analysis is also the natural setting for many other problems in engineering, mathematics, and the sciences. For example, Wiener's Tauberian theorem in

Fourier analysis not only characterizes the behavior of the prime numbers, but also provides the proper notion of spectrum for phenomena such as white light; this latter process leads to the Fourier analysis associated with correlation functions in filtering and prediction problems, and these problems, in turn, deal naturally with Hardy spaces in the theory of complex variables.

Nowadays, some of the theory of PDEs has given way to the study of Fourier integral operators. Problems in antenna theory are studied in terms of unimodular trigonometric polynomials. Applications of Fourier analysis abound in signal processing, whether with the fast Fourier transform (FFT), or filter design, or the adaptive modeling inherent in time-frequencyscale methods such as wavelet theory.

The coherent states of mathematical physics are translated and modulated Fourier transforms, and these are used, in conjunction with the uncertainty principle, for dealing with signal reconstruction in communications theory. We are back to the raison d'^etre of the *ANHA* series!

University of Maryland   John J. Benedetto
College Park   Series Editor

Preface

While preparing this volume, I was encouraged to present the details of three new claims that I have made concerning Morrey Spaces and the action of Riesz potential operators on these spaces. The claims all follow from a new formulation of the predual to a Morrey Space (2012): $H^{p',\lambda}(\mathbb{R}^n)$; see Chapter 5, a version that generally meets all the desired requirements for such in Harmonic Analysis. The full theory is presented here for the first time. These include:

(1) Determine the integrability classes of the trace of a Riesz potential of an arbitrary Morrey function, in the unbounded case. The key idea here is the existence of certain Wolff potentials; see Chapter 10.

(2) Determine the capacity of the singular sets (sets of discontinuities) of weak solutions of a $2m$th-order quasilinear elliptic systems, with mth-order derivatives belonging to L^p, e.g., the celebrated Meyers-Elcrat system. Here the mth-order derivatives can be upgraded to belong to a Morrey class; see Chapters 15 and 16.

(3) Are there any "full" interpolation results for linear operators between Morrey Spaces in the light of the now known counterexamples in the literature? I claim that there are - with important restrictions. For this we refer the reader to Chapter 11. This all depends on an atomic decomposition of a Morrey Space and a full use of our duality of Chapter 5.

Lexington, KY, USA   David R. Adams
July 2014

Contents

<u>Lectures</u>
<u>on</u>
<u>Morrey Spaces</u>
by
David R. Adams
2013

Soaring thru Morrey Space
Just below the Lebesgue clouds
And avoiding the sharp Holder rocks below,
I see more clearly now
The Hausdorff dimensions of our mathematical coastlines
I am a Riesz Potential Operator
and I am free!

Chapter 1
Introduction

In these lecture notes, I intend to discuss the theory of Morrey Spaces, whose development recently has placed them solidly in the mainstream of modern Harmonic Analysis (HA), defined here by the works of Stein and Torchinsky; [St2] and [To]. This has been accomplished recently by the work [AX2 - 5] and the work of others as noted throughout these notes. These spaces were introduced by C. B. Morrey in 1938 [Mo] in his work on systems of second order elliptic partial differential equations (PDE) and together with the now well-studied Sobolev Spaces, constitute a formidable three parameter family of spaces useful for proving regularity results for solutions to various PDE, especially for non-linear elliptic systems.

We will say that a function $f \in L^{p,\lambda} = L^{p,\lambda}(\mathbb{R}^n)$ if

$$\sup_{\substack{r > 0 \\ x_0 \in \mathbb{R}^n}} \left[r^{\lambda-n} \int_{B(x_0,r)} |f(y)|^p \, dy \right]^{1/p} < \infty. \tag{1.1}$$

Here, $1 < p < \infty$, $0 \leq \lambda \leq n$, and the quantity of (1.1) is the (p, λ) - Morrey norm, denoted by $\|f\|_{L^{p,\lambda}}$. One can also define the corresponding Morrey Space over a domain Ω contained in $\mathbb{R}^n$ by just integrating in (1.1) over the sets $\Omega \cap B(x_0, r)$ and taking the supremum over only $0 < r \leq \text{diam } \Omega = r_o$. This is usually done when the supp $f \subset \Omega$, and Ω is a domain of "type A" in the sense of Troianiello [Tr] : $|\Omega \cap B(x_0, r)| \geq A r^n$, for all $x_0 \in \Omega$ and $r \leq \text{diam } \Omega$. The main reason for this condition is that one can trivially then extend a function f in Ω to all of $\mathbb{R}^n$ - set $f \equiv 0$ outside Ω- and conclude that the extended function belongs to $L^{p,\lambda}$. Here we use the notation $L^{p,\lambda}(\Omega)$ for such a "local" version of $L^{p,\lambda}$. But mostly, we will deal with Morrey Spaces over the entire $\mathbb{R}^n$ (and in Chapter 17 on non-compact Riemannian manifolds).

© Springer International Publishing Switzerland 2015
D. Adams, *Morrey Spaces*, Applied and Numerical Harmonic Analysis,
DOI 10.1007/978-3-319-26681-7_1

Notice that $L^{p,n} \equiv L^p$, the usual Lebesgue space of the pth power integrable functions on $\mathbb{R}^n$ (or on Ω in the local version), and $L^{p,0} \equiv L^\infty$. However, be it noted here, that throughout, we will be interested in the case $0 < \lambda < n$, especially in Chapter 11, because we will always need: $0 < n - \lambda$.

Also, we will use the case $p = 1$ in (1.1) when the f is replaced by signed (Borel) measures on $\mathbb{R}^n$, i.e.

$$\sup_{\substack{r > 0 \\ x_0 \in \mathbb{R}^n}} \left(r^{\lambda-n} \int_{B(x_0,r)} |d\mu(y)| \right) = ||\mu||_{L^{1,\lambda}} < \infty. \tag{1.2}$$

Such measures arise, for example, in Frostman's Theorem (see Section 3.3).

Surely the main fame associated with Morrey Spaces is the celebrated "Morrey's Lemma," and many results in mathematical analysis rely on exactly this concept of the growth/decay of an integral average of a function over a ball; see in particular Moser's iteration argument leading to the Harnack principle for PDE [BB].

Morrey's Lemma: If $|\nabla u| \in L^{p,\lambda}$ with $p > \lambda$, then $u(x) \in C^\alpha(\mathbb{R}^n)$, $\qquad$ (1.3)

the space of Hölder continuous functions on $\mathbb{R}^n$, with exponent $\alpha = 1 - \lambda/p$.

Often this lemma is the gateway to higher order regularity for solutions of elliptic PDE.

The use of the term "regular" in these notes generally denotes smoothness in the designated neighborhood, even C^∞ there; see [BF].

In the 1960s, S. Campanato defined a scale of spaces, $\mathscr{L}_k^{p,\lambda}$, that included the distinguished quartet:

$$L^p - L^{p,\lambda} - \mathrm{BMO} - C^\alpha, \tag{1.4}$$

at least locally. His definition used the mean oscillation of the function f, i.e.

$$\sup_{x,r>0} \left[r^{\lambda-n} \int_{B(x,r)} |f(y) - P(y)|^p \, dy \right]^{1/p} < \infty \tag{1.5}$$

with $B(x, r)$ a ball in $\mathbb{R}^n$ centered at x and of radius r, and $P(y)$ is a polynomial of degree $\leq k$. But in these notes, we will always take $k = 0$ and the corresponding constant $= f_r(x) =$ integral average of f over the ball $B(x, r)$. Then we will take only

$$-p < \lambda < n$$

throughout, and get

$$\mathscr{L}_0^{p,\lambda}(\mathbb{R}^n) = \begin{cases} L_{loc}^{p,\lambda}, & 0 < \lambda < n \\ \text{BMO}, & \lambda = 0 \\ C^\alpha, & -p < \lambda < 0, \alpha = -\lambda/p. \end{cases} \tag{1.6}$$

We will discuss this arrangement more fully in Chapter 2. These results have been nicely summarized in the paper by J. Peetre [P], upto 1969. For more recent sources, see, for example, [Tr], 1987. The subscript $+$ on a function space denotes the non-negative elements in that function space; e.g., $L_+^{p,\lambda}$.

It is probably because of (1.6) that these spaces gained considerable popularity for regularity considerations for elliptic PDE in the late 1960s and early 1970s, especially because of the inclusion of BMO (the space of functions of bounded mean oscillation of John and Nirenberg). And no doubt this was enhanced by the discovery of the predual of BMO by C. Fefferman in the early 1970s: the real Hardy Space $H^1(\mathbb{R}^n)$. The reader is referred to the excellent treatments found in [St2] and [To].

But in recent years, the excitement over these and Morrey type spaces has somewhat dissipated, save for many attempts to generalize the Morrey condition via growth/decay of various averages. One reason for this, perhaps, is that either the predual of $L^{p,\lambda}$ was not known or the version had its drawbacks, perhaps even unusable in the sense that its ties to modern Harmonic Analysis were quite weak, again as defined by [St2] or [To]. Indeed, some attempts at a predual were given as early as 1986: [A5, Z, K], and most recently in [AX2]. But the author believes that it wasn't until [AX3] that a successful version of the predual to $L^{p,\lambda}$ was given - that meets all the needs of HA. This space is denoted in these notes by $H^{p',\lambda}(\mathbb{R}^n)$, or just $H^{p',\lambda}$, $p' = p/(p-1)$. And the real interesting feature here is that its definition relies on the notion of the Choquet Integral with respect to Hausdorff capacity on $\mathbb{R}^n$ and the special class of Muckenhoupt A_1 weights. This theory is discussed in Chapters 4 and 5, after a review of Hausdorff Capacity $\Lambda^d(\cdot)$ of dimension d, $0 < d \le n$, and its more well-known counterpart, Hausdorff measure $\mathcal{H}^d$; and n - dimensional Lebesgue measure on $\mathbb{R}^n$, $\mathscr{L}^n$.

$$H^{p,\lambda} = H^{p,\lambda}(\mathbb{R}^n)$$

$$= \left\{ f \in L_{loc}^p : \inf_\omega \int |f(y)|^p \omega(y)^{1-p} dy < \infty \right\} \tag{1.7}$$

where the infimum is over all non-negative weights w satisfying

$$\omega \in A_1 \text{ and } \int \omega \, d\Lambda^{n-\lambda} \le 1. \tag{1.8}$$

Here A_1 denotes

$$\fint_Q \omega \, dx \le c \cdot \operatorname*{ess\,inf}_Q \omega$$

for some constant c independent of Q = cube in $\mathbb{R}^n$ with sides parallel to the coordinate axes; f_Q = integral average over Q. So in the language of Functional Analysis:

$$(H^{p',\lambda})^* = L^{p,\lambda}. \tag{1.9}$$

A further history of Morrey Spaces has been the demonstration that the classical operators of HA are indeed bounded on $L^{p,\lambda}$: the Calderon-Zygmund singular integrals (Chapter 8), the maximal operators (Chapter 6), and the potential operators (Chapter 7); see [P, A3, CF] and others. But with the new predual result (1.9), all of these results can be made mere corollaries to the now classical theory of A_p weights in HA.

It should also be mentioned that with this new predual formulation, one can develop a interpolation result for such operators between Morrey Spaces (Chapter 11), provided the Morrey parameter λ stays strictly in the range: $0 < \lambda < n$. Previous attempts, e.g., by Stampacchia [S] and Peetre [P] were limited to the case when Morrey Spaces were in the ranges of the operators with L^p domains, not $L^{p,\lambda}$. Also, a recent result of Yang-Yuan-Zhuo [Y²Z] gives a result that implies that if $L^{p,\lambda}$ is allowed to be in the domain $0 < \lambda \le n$, then they need $\lambda_0 = \lambda_1$, i.e., the λ parameter must be fixed for their proof to work. Furthermore, the full interpolation for $0 < \lambda_0 \le \lambda_\theta \le \lambda_1 \le n$ is clearly false when λ_1 is allowed to be equal to n. Such is given in the paper by Blasco-Ruiz-Vega [BRV]. So it is natural to surmise from this that one needs $\lambda_0 = \lambda_1$, as the positive result in [Y²Z] states. However, Theorem 11.2 states otherwise! But one of course must always have $\lambda_1 < n$, a very natural condition imposed by the very definition of $H^{p,\lambda}$; i.e., there is no role for $\Lambda^{n-\lambda_1}, \lambda_1 = n$ in these deliberations. More will be said below in Chapter 11. Also, it should be noted that the example given in [SZ] is also of this type.

In further chapters, we give application of the Morrey Theory to: Morrey-Sobolev Capacities (Chapter 9), trace of Morrey potentials with the multiplier operator (Chapter 10). These together with the last three chapters reflect the interest of the author and by no means exhaust applications. However, the author is quite partial to the presentations given in Chapters 15 and 16 on non-linear elliptic PDE and their exceptional sets.

As it turns out, the Morrey condition on the derivatives of such solutions (solutions to certain non-linear elliptic PDE) can give a "regularity condition" on the solutions even when one operates below the continuity threshold, i.e., in the language of (1.3), when $p < \lambda$. The basic result there is: the singular set (= the set there the solution u is discontinuous) must have Hausdorff dimension $\le n - p$. These lecture notes are an attempt to extend such a theory; [G].

Such non-linear elliptic systems include the celebrated Meyers-Elcrat system (see [ME] and [BF]):

$$\sum_{|\gamma| \le m} (-1)^\gamma \left(\frac{\partial}{\partial x} \right)^\gamma A_\gamma(x, D^m u) = 0, \text{ in } \Omega, \tag{1.10}$$

with the now refined estimate on the dimension of the local exceptional sets = singular sets in Ω. Examples (first due to De Giorgi) in the late 1960s show that such singularities can occur for such systems, but by the famous results of De Giorgi and Nash, cannot occur for single divergent form elliptic equations (with bounded measurable coefficients). Other subsequent examples basically refine De Giorgi's theory; the reader is referred to the excellent book [BF]. Most of the Sobolev theory of such singular sets is contained in the treatment of Giaquinta [G]. The use of Morrey-Sobolev capacities refines Giaquinta's work: the passage from the Sobolev regime to the Morrey-Sobolev regime.

Also, some mention should be made of the extension of the Morrey-Sobolev function theory to the Besov scale. This scale is of course treated extensively in $[Y^2Z]$, but for some time the analogue of the Morrey-Sobolev inequality (first established in [A3]) has been proven in the Morrey-Besov setting in [AL]- in 1975 for the former and 1982 for the latter. These two inequalities play the role of the Sobolev inequality in their respective settings; Theorem 7.1(i) vs Theorem 14.1 of Chapters 7 and 14.

And finally, we come to Chapter 17 where we prove and/or speculate on the extension of earlier results to the setting of non-compact complete Riemannian manifolds (CRM) of dimension $n \geq 3$. The main idea here is to learn enough to understand the Sobolev inequalities/embeddings in the CRM setting-results that go back mainly to Varopoulos in the 1980s, but are here taken from the recent works of Saloff-Coste and Hebey (1990s). Once we have this under our belt, the next thing is to propose a proof of an analogue of the Morrey-Sobolev inequality for potentials of Morrey functions on CRM with the basic: maximal volume growth and Ricci curvature bounded below; see Section 17.1. But again the reader is invited to improve and extend these results and resolve some of the open conjectures.

The reader can view these notes as a new enhanced version of the theory of Morrey Spaces now in fragments in the literature. However, these notes constitute a work in progress and should be treated as such.

Our policy on proofs will generally be: if it is a result well established in [AH, St2], or [To], then the coordinates of such will be given. But also we will rely heavily on the papers [AX2, AX3, AX4, AX5, AX6, AX7], with most of these arguments presented/modified here in. Occasionally, just an outline of a proof will be given, but enough to convince the reader it is true - or he/she should take up pen and set the author right!

But the key word is to: Enjoy!

David R Adams
July 2014.
(2nd ed)

These notes were typed from a hand written version by Hem Raj Joshi, Associate Professor, Department of Mathematics and CS, Xavier University, Cincinnati, Ohio.

Some Morrey Space Time Lines

1938	C. B. Morrey [Mo]: Morrey's Lemma;
1953	G. Choquet [Ch]: Choquet integrals;
1963	S. Campanato [Ca] : $\mathscr{L}_k^{p,\lambda}$ scale;
1965	G. Stampacchia [S]: Interpolation and weak type embedding;
1969	J. Peetre [P]: Singular integrals on Morrey Spaces;
1975	D. R. Adams [A1]: Strong type embedding (potential operator);
1986	D. R. Adams [A5]: Predual $p = 1$;
1986	C. Zorko [Z]: Atomic predual for $\mathscr{L}_0^{p,\lambda}$;
1987	F. Chiarenza-M. Frasca [CF]: Maximal Operator on Morrey Spaces;
1998	E. Kalita [K]: A predual for $L^{p,\lambda}$ especially for PDE;
2004	D. R.Adams-J. Xiao [AX2]: Capacity, preduality and atomic decomposition for $L^{p,\lambda}$;
2010	D. R.Adams-J. Xiao [AX3]: Preduality, Interpolation, and Wolff potentials;
2011	D. R.Adams-J. Xiao [AX4]: Traces of Morrey Potentials;
2012	D. R.Adams-J. Xiao [AX5]: Commutators;
2012	D. R.Adams-J. Xiao [AX6, AX7]: Appl. to PDE;
2003–2012	V. Buenkov-V. Guliyeu, M. Dzhabrailov- S. Khaligova, V. Guliyev, A. Mazzucato, Z. Shen [BG, DK, Gu, Maz, Sh]: Generalized Morrey Spaces with applications.

Chapter 2
Function Spaces

2.1 $L^p, C^\alpha, \text{BMO}, L^{p,\lambda}, L_w^{p,\lambda}$

Listed here are several classical function spaces defined with some of their
properties useful in the sequel. Generally, Ω will be a bounded domain in $\mathbb{R}^n$, and
when it matters, with smooth boundary $\partial\Omega$ or at least a boundary of type A (as noted
in the Introduction). L^p denotes the usual Lebesgue space of pth power integrable
functions on $\mathbb{R}^n$ or respectively on $\Omega, L^p(\Omega); \ 1 \leq p < \infty$, and $||f||_{L^p}$ or $||f||_{L^p(\Omega)}$.
On the other hand, weak - $L^p(L_w^p)$ consists of those f for which

$$\sup_{t>0}\left[t^p \, \mathscr{L}_n \left(\{x \in \mathbb{R}^n : |f(x)| > t\}\right)\right]^{1/p} \equiv ||f||_{L_w^p} < \infty,$$

and the corresponding $L_w^p(\Omega). L^\infty$ is the essentially bounded functions on $R^n, L^\infty(\Omega)$
those on Ω. Whereas, C^α is the usual space of Hölder continuous functions on $\mathbb{R}^n$
with exponent $\alpha \in (0, 1)$ and normed by

$$||f||_{C^\alpha} = ||f||_{L^\infty} + [f]_\alpha < \infty,$$

where,

$$[f]_\alpha = \sup_{\substack{x, y \in \mathbb{R}^n \\ x \neq y}} \frac{|f(x) - f(y)|}{|x - y|^\alpha},$$

and the corresponding $C^\alpha(\Omega)$. We shall also have some need of the John-Nirenberg
space of functions of bounded mean oscillation on a fixed cube $Q_0 \subset \mathbb{R}^n$ or on $\mathbb{R}^n$
itself, i.e., $f \in \text{BMO}(Q_0)$ if

© Springer International Publishing Switzerland 2015
D. Adams, *Morrey Spaces*, Applied and Numerical Harmonic Analysis,
DOI 10.1007/978-3-319-26681-7_2

$$\sup_{Q \subset Q_0} \int_Q |f(x) - f_Q| dx = [f]_{*,Q_0} < \infty,$$

where Q is a cube in $\mathbb{R}^n$ with sides parallel to the coordinate axes; f_Q = the integral average of f over Q. For BMO on $\mathbb{R}^n$, just replace Q_0 by $\mathbb{R}^n$. And in particular, we have the celebrated J-N Lemma:

Lemma 2.1. If $[f]_{*,Q_0} < \infty$, then there exist two constants c_1 and c_2 such that

$$\mathscr{L}_n(\{x \in Q : |f(x) - f_Q| > \lambda\}) \leq c_1 e^{-c_2 \lambda /[f]_{*,Q_0}} \cdot |Q|$$

for all cubes $Q \subset Q_0$ and any $\lambda > 0$. Here $|Q| = \mathscr{L}_n(Q)$.

(For this, see either [St2] or [To]). It thus follows that $L^\infty(Q_0) \subset \mathrm{BMO}(Q_0) \subset L^p(Q_0)$ for all $p \geq 1$.

2.2   The Campanato scale $\mathscr{L}^{p,\lambda}$, $\mathscr{L}^{p,\lambda}(Q_0)$

For $-p < \lambda \leq n$, set

$$\mathscr{L}^{p,\lambda}(Q_0) = \{f : ||f||_{L^p(Q_0)} + [f]_{p,\lambda;Q_0} < \infty\}$$

where

$$[f]_{p,\lambda;Q_0} = \sup_{Q \subset Q_0} \left(|Q|^{-\lambda/n} \int_Q |f - f_Q|^p \, dx \right)^{1/p}.$$

As mentioned, this scale includes many of the classical function spaces of Harmonic Analysis, notably $L^p(Q_0), L^{p,\lambda}(Q_0), \mathrm{BMO}(Q_0),$ and $C^\alpha(Q_0)$ - as well as versions over $\mathbb{R}^n$. We will not give a complete proof of (1.6) even just over Q_0, but refer the reader to [Tr] for missing details. But because these notes are about Morrey Spaces, it seems appropriate to at least prove:

$$\mathscr{L}_0^{p,\lambda}(Q_0) = L^{p,\lambda}(Q_0) \tag{2.1}$$

with equivalence of norms: $0 < \lambda \leq n, \ 1 < p < \infty$. In fact this is a consequence of the following iteration lemma:

Lemma 2.2. Let $\varphi(r)$ be a non-negative function on $(0, R]$ and suppose there are numbers $\beta, \gamma > 0$ and $K > 1$ such that

$$\varphi(\rho) \leq K \left(\frac{\rho}{r} \right)^\beta \varphi(r) + K\rho^\gamma$$

whenever $\frac{r}{s} \le \rho < r$ and $r \le R$, for some $s > 1$ and $R < \infty$. Then for any $0 < \epsilon < \beta - r$,

$$\varphi(\rho) \le K\left(\frac{\rho}{r}\right)^{\beta-\epsilon} \varphi(r) + KC\rho^\gamma \tag{2.2}$$

for some constant C depending only on K, β, γ. In particular, (2.2) holds for all $0 < \rho < r = R$. This Lemma is a special case of Lemma 1.18 of [Tr].

Applying this lemma to the estimate below gives the inclusion $\mathscr{L}^{p,\lambda}(Q_0) \subset L^{p,\lambda}(Q_0)$. The reverse is obvious. So let $Q = Q_\rho$ = a cube with edge length ρ, then

$$\varphi(\rho) = \int_{Q_\rho} |u|^p \le 2^p \int_Q |u - u_Q|^p + 2^p \int_Q |u_Q|^p$$

$$\le 2^p \rho^{n-\lambda} [u]^p_{p,\lambda;Q_0} + 2^p \frac{\rho^n}{r^n} \int_{Q_r} |u|^p.$$

Then with Lemma 2.2 and $\gamma = n - \lambda < n - \epsilon$ for some $\epsilon > 0$, and $\beta = n$, it follows that

$$\varphi(\rho) \le C\rho^{n-\lambda}\left([u]^p_{p,\lambda;Q_0} + \|u\|^p_{L^p(Q_0)}\right)$$

for all $0 < \rho < R$.

For the case $\lambda = 0$, applying the J-N Lemma gives $1 < p < \infty$

$$\text{BMO}(Q_0) = \left\{f : \sup_{Q \subset Q_0} \left(\fint_Q |f - f_Q|^p \, dx\right)^{1/p} < \infty\right\},$$

for any $Q_0 \subset \mathbb{R}^n$.

Finally, when $-p < \lambda < 0$, we remark that

$$\mathscr{L}^{p,\lambda}(Q_0) \equiv C^\alpha(Q_0), \alpha \equiv -\lambda/p > 0, \tag{2.3}$$

was the consequence of independent investigations by S. Campanato [Ca] and N.G. Meyers [M2]. See [To], VIII.5. The proof in this case is a bit more delicate than that given above for $0 < \lambda \le n$, but is repeated in several sources, namely in both [Tr] and [G].

2.3   Sobolev Spaces $W^{m,p}(\Omega), G_\alpha(L^p), I_\alpha(L^p)$

The space $W^{m,p}(\Omega)$ consists of those weakly differentiable functions $u(x)$ on Ω for which

$$\int_\Omega |D^m u|^p \, dx + \int_\Omega |u|^p \, dx \equiv ||u||^p_{W^{m,p}(\Omega)} < \infty,$$

where m = positive integer, $D^m u$ = the set of all m^{th} order derivatives of u on Ω. But for us, it will be essential to have a potential theoretic version of $W^{m,p}(\mathbb{R}^n)$. Here we refer to [AH] with $G_\alpha(x)$ the Bessel potential operator, $\alpha > 0$, $\hat{G}_\alpha(\zeta)$ = (Fourier transform of G_α) = $(1 + |\zeta|^2)^{-\alpha/2}$, $\zeta \in \mathbb{R}^n$. So when $\alpha = m$, a positive integer, then $W^{m,p}(\mathbb{R}^n) = G_\alpha(L^p(\mathbb{R}^n))$, and with $u(x) = G_m f$, $||u||_{W^{m,p}(\mathbb{R}^n)} \sim ||f||_{L^p(\mathbb{R}^n)}$, with equivalent norms; $1 < p < \infty$.

Now if $I_\alpha(x) = |x|^{\alpha-n}$, $0 < \alpha < n$, and $f \in L^p(\mathbb{R}^n)$, $I_\alpha f(x) = \int_{\mathbb{R}^n} |x - y|^{\alpha-n} f(y) \, dy$, the Riesz potential of f of order α (which is finite a.e. when $\alpha p < n$), then

$$||f||_{L^p(\mathbb{R}^n)} + ||I_\alpha f||_{L^p(\mathbb{R}^n)}$$

is an equivalent Sobolev norm, $\alpha = m < n, 1 < p < n/\alpha$. This is the result of the fact that the Calderon-Zymund singular integrals are bounded on $L^p(\mathbb{R}^n)$, $1 < p < \infty$. See [St2] and [To].

2.4   Morrey-Sobolev Spaces $I_\alpha(L^{p,\lambda})$

From the above, we shall denote by $I_\alpha(L^{p,\lambda}(\mathbb{R}^n))$ a Morrey-Sobolev space, $0 < \lambda < n$, $1 < p < n/\alpha$, $0 < \alpha < n$. And often we will write $I_\alpha f$ when f has compact support, and then from the results latter of Chapter 8, we set

$$||u||_{W^{m,p;\lambda}} = ||u||_{L^{p,\lambda}(\mathbb{R}^n)} + ||f||_{L^{p,\lambda}(\mathbb{R}^n)},$$

when $u(x) = I_m f$; i.e., u and its m-th order derivatives belong to the Morrey Space $L^{p,\lambda}(\mathbb{R}^n)$.

2.5   Dense/non-dense subspaces, Zorko Spaces, $VL^{p,\lambda}$, VMO

It is well known that class $C_0^\infty(\Omega) = C^\infty$ functions on Ω with compact support in Ω is dense in $W^{m,p}$ and in $G_\alpha(L^p)$ when $\Omega = \mathbb{R}^n$. What we seek here is the density subspaces for $L^{p,\lambda}(\Omega)$. For this we turn to Zorko [Z].

Our first observation is: there are $f \in L^{p,\lambda}$ that cannot be approximated even by continuous functions in the norm $|| \cdot ||_{L^p} + || \cdot ||_{L^{p,\lambda}}$. In fact Zorko shows that $f_{x_0}(x) = |x - x_0|^{-\lambda/p}$, $x_0 \in \Omega$, is one, for every x_0. In fact

$$\int_{|x-x_0|<\rho} |f_{x_0}(x) - g(x)|^p dx \geq 2^{-p} \int_{|x-x_0|<\rho} |f_{x_0}(x)|^p \, dx - \int_{|x-x_0|<\rho} |g(x)|^p \, dx$$

$$\geq \frac{2^{-p} w_{n-1}}{n-\lambda} \rho^{n-\lambda} - ||g||_{L^\infty(B(x_0,\rho))} \cdot \frac{w_{n-1}}{n} \rho^n$$

$$= w_{n-1} \rho^{n-\lambda} \left(\frac{2^p}{n-\lambda} - ||g||_{L^\infty(B(x_0,\rho))}^p \cdot \rho^\lambda \right).$$

So for $\rho \leq \rho_0 =$ sufficiently small,

$$\rho^{\lambda-n} \int_{|x-x_0|<\rho} |f_{x_0}(x) - g(x)|^p dx \geq c_0 > 0.$$

This motivates us to set

$$VL^{p,\lambda}(\mathbb{R}^n) = \left\{ f \in L^{p,\lambda} : ||f(\cdot - y) - f(\cdot)||_{L^{p,\lambda}} \to 0, \ |y| \to 0 \right\}.$$

Here the V stands for "vanishing" honoring Sarason's VMO = Vanishing mean oscillation function space; see [To]. We also at times will refer to this space as the Zorko subspace of $L^{p,\lambda}$, the Morrey Space on $\mathbb{R}^n$. We now note

Theorem 2.3. If $f \in VL^{p,\lambda}$, then f can be approximated by $C_0^\infty(\mathbb{R}^n)$ in the norm

$$|| \cdot ||_{\mathscr{L}^{p,\lambda}} = || \cdot ||_{L^p(\mathbb{R}^n)} + || \cdot ||_{L^{p,\lambda}(\mathbb{R}^n)}.$$

Proof. Let $\varphi \in C_0^\infty(B(0,1))^+$ with $\int \varphi(x)dx = 1$, then upon setting $\varphi_\epsilon(x) = \epsilon^{-n}\varphi(x/\epsilon)$, $\epsilon > 0$, and $\varphi_\epsilon * f$ the usual convolution, we get

$$\left(\int_{B(x_0,\rho)} |\varphi_\epsilon * f - f|^p \, dx \right)^{1/p} \leq \int \varphi_\epsilon(y) \left\{ \int_{B(x_0,\rho)} |f(x-y) - f(x)|^p \, dx \right\}^{1/p} dy$$

hence

$$\left(\rho^{\lambda-n} \int_{B(x_0,\rho)} |\varphi_\epsilon * f - f|^p \, dx \right)^{1/p} \leq \int \varphi_\epsilon(y) ||f(\cdot - y) - f(\cdot)||_{L^{p,\lambda}} \, dy$$

or

$$||\varphi_\epsilon * f - f||_{L^{p,\lambda}} \leq \sup_{|y|<\epsilon} ||f(\cdot - y) - f(\cdot)||_{L^{p,\lambda}}.$$

$\square$

This may seem a bit disconcerting, but as it turns out, all is not lost, for we have:

Theorem 2.4. If $f \in L^{p,\lambda}$, then $f \in VL^{p,\mu}$ for all $\mu > \lambda$, $0 < \lambda < n$.

Proof. Again by mollifying f as above, we can write

$$r^{\mu-n} \int_{B(x_0,r)} |f - f * \varphi_\epsilon|^p$$

$$\leq \left(r^{\lambda-n} \int_{B(x_0,r)} |f - f * \varphi_\epsilon|^p \, dx \right)^{1/q} \left(\int_{B(x_0,r)} |f - f * \varphi_\epsilon|^p \, dx \right)^{1/q'}$$

where $\lambda < \mu < n$, $q = (n - \lambda)/(n - \mu)$, and $q' = q/(q - 1)$. The first factor is bounded by $C \|f\|_{L^{p,\lambda}}$ and the second factor tends to zero as $\epsilon \to 0$ due to the density of smooth functions in the L^p spaces. $\qquad\square$

2.6 Note

1. Extensions of the idea of a Morrey Space and the study of various operators (classical or not) on these spaces are numerous. We collect a few of these that have caught our attention in the Bibliography under "Generalized Morrey Spaces and some applications."

Chapter 3
Hausdorff Capacity

3.1 Set functions Λ^d, $\mathcal{H}^d$, $\mathscr{L}^n$, $0 < d \leq n$:

Given a number $d \in (o, n]$, the d-dimensional Hausdorff Capacity of a set $E \subset \mathbb{R}^n$ is given by

$$\Lambda^d(E) = \inf \left\{ \sum_j r_j^d : E \subset \cup_j B(x_j, r_j), \ j = 1, 2, 3, \dots \right\} \tag{3.1}$$

i.e., E is covered by balls $B(x_j, r_j)$, centered at some x_j and of radius $r_j > 0$, and then the infimum is taken over all corresponding sums. If one restricts the radii of the covering balls to satisfy: $r_j \leq \epsilon$, for some $\epsilon > 0$, then we would write $\mathcal{H}^d_{(\epsilon)}(E)$ for the infimum, and $\mathcal{H}^d(E)$ for the corresponding limit of $\mathcal{H}^d_{(\epsilon)}(E)$ as $\epsilon \to 0$ (such a limit always exists as possibly $+\infty$ - for most sets $E \subset \mathbb{R}^n$). However, $\Lambda^d(E)$ is always finite whenever E is bounded. Λ^d is also called d-dimensional Hausdorff content by several authors. We use the term capacity because Λ^d is indeed a capacity set function in the sense of N. G. Meyers [M2].

$$\underline{C \text{ is a capacity in the sense of N. G. Meyers if}} \tag{3.2}$$

  i. C is a non-negative set function on all subsets of $\mathbb{R}^n$:
 ii. C is monotone non-decreasing;
      $C(E) \leq C(F)$, $E \subset F$;
iii. C is countable subadditive:
      $C(\cup_j E_j) \leq \sum_j C(E_j)$;
 iv. $C(\phi) = 0$, $\phi = $ empty set.

Further useful properties of some capacities are described by Choquet [C]:

© Springer International Publishing Switzerland 2015
D. Adams, *Morrey Spaces*, Applied and Numerical Harmonic Analysis,
DOI 10.1007/978-3-319-26681-7_3

v. for any increasing sequence of sets E_j with $E = \cup_j E_j$, $E_j \subset E_{j+1}$,
   then $C(E) = \lim_{j \to \infty} C(E_j)$

vi. for any decreasing sequence of compact sets K_j, with $K = \cap_j K_j$, $K_j \supset K_{j+1}$,
   then $C(K) = \lim_{j \to \infty} C(K_j)$.

Capacities that satisfy (i)–(vi) are called <u>Choquet</u> <u>Capacities</u>.

3.2 Dyadic versions: $\tilde{\Lambda}^d$, $\tilde{\Lambda}^d_0$

Definition 3.1. The dyadic cubes in $\mathbb{R}^n$ are: let $\mathbb{Q}_0$ be a collection of cubes in $\mathbb{R}^n$ which are congruent to $(0, 1]^n$ and whose vertices lie on the lattice $\mathbb{Z}^n$. Dilating this family by a factor 2^{-k} yields the collection we will denote by $\mathbb{Q}_k$, $k \in \mathbb{Z}$. Note that each cube in this collection is "half open" and has vertices at points $(2^{-k}\mathbb{Z})^n$. This collection will be referred to as the "dyadic cubes of $\mathbb{R}^n$".

<u>Dyadic Hausdorff Capacity</u> We set

$$\tilde{\Lambda}^d(E) = \inf\left\{ \sum_j l(Q_j)^d : E \subset \cup_j Q_j, \; Q_j \text{ dyadic}\right\} \qquad (3.3)$$

where $l(Q) = $ edge length of Q.

However, below we shall have need for a restricted version of Hausdorff Capacity, namely the Dyadic Interior Hausdorff Capacity. In this sequel we follow closely the work of [Y²].

<u>Dyadic Interior Hausdorff Capacity</u>

$$\tilde{\Lambda}^d_0(E) = \inf\left\{ \sum_j l(Q_j)^d : E \subset (\cup_j Q_j)^o, \; Q_j \text{dyadic}\right\}, \qquad (3.4)$$

where $E^O = $ interior of E.

First of all notice that

$$\tilde{\Lambda}^d_0(E) \le \tilde{\Lambda}^d(E) \le \Lambda^d(E) \le \mathcal{H}^d(E) \qquad (3.5)$$

for all subsets $E \subset \mathbb{R}^n$, and that Λ^d and $\mathcal{H}^d$ have the same null sets. (Indeed, if $\Lambda^d(E) = 0$, then there is a cover of E by balls such that $\sum_j r_j{}^d < \eta$, for any $\eta > 0$.

But this implies that $r_j \le \eta^{1/d}$ and thus $\mathcal{H}^d_{(\epsilon)}(E) \le \eta$, whenever $\eta^{1/d} < \epsilon$. So $\mathcal{H}^d(E) = 0$).

3.3 Frostman's Theorem

Our first result here is the classical theorem of Frostman; see[AH].

Theorem 3.1. Let K be a compact subset of $\mathbb{R}^n$, then $\Lambda^d(K) > 0$ iff there exists a Borel measure μ supported on K such that

$$\mu(B(x,r)) \leq Ar^d \tag{3.6}$$

for all x and all $r > 0$; here A is a constant independent of x and r. Furthermore, there is a constant $a > 0$ independent of K such that

$$\frac{1}{a}\Lambda^d(K) \leq \mu(K) \leq a \cdot \Lambda^d(K). \tag{3.7}$$

Proof. We refer the reader to [AH] page 136.   □

Next, we develop the further relationships between the various Hausdorff capacities.

Theorem 3.2. For any subset $E \subset \mathbb{R}^n$,

$$\frac{1}{6^n}\tilde{\Lambda}_0^d(E) \leq \Lambda^d(E) \leq n^n \cdot \tilde{\Lambda}_0^d(E); \tag{3.8}$$

i.e., $\tilde{\Lambda}_0^d$ as a set function is equivalent to Λ^d in this rough sense.

Proof. First for the right hand inequality: Let $\{Q_j\}$ be a dyadic cover of E such that $E \subset (\cup_j Q_j)^o$. Then each Q_j can be covered by at most n^n balls of radius $l(Q_j)$. We call these balls $B_{j,s}, s = 1, \cdots, n^n$. So

$$\Lambda^d(E) \leq \sum_j \sum_{s=1}^{n^n} r_{B_{j,s}}{}^d = n^n \sum_j l(Q_j)^d.$$

For the left hand inequality: Cover E by open balls with radius r_j. Then there exists integers $k_j < 0$ such that $2^{k_j} \leq r_j < 2^{k_j+1}$ and then each $B(x_j, r_j)$ can be covered by at most 6^n dyadic cubes with edge length $l = 2^{k_j}$, $\{Q_{j,s}\}$, $s = 1, \cdots, 6^n$, and

$$E \subset \cup_j B(x_j, r_j) \subset \cup_j(\cup_{s=1}^{6^n} Q_{j,s})^0 \subset (\cup_j \cup_{i=1}^{6^n} Q_{j,s})^0$$

Hence

$$\tilde{\Lambda}_0^d(E) \leq 6^n \sum_j r_j^d.$$

□

3.4   Strong subadditivity of $\tilde{\Lambda}_0^d$ and $\Lambda^d \sim \tilde{\Lambda}_0^d$

Theorem 3.3. The capacities $\tilde{\Lambda}_0^d(E)$ are strongly subadditive, i.e.

$$\tilde{\Lambda}_0^d(E_1 \cup E_2) + \tilde{\Lambda}_0^d(E_1 \cap E_2) \le \tilde{\Lambda}_0^d(E_1) + \tilde{\Lambda}_0^d(E_2), \qquad (3.9)$$

for any sets $E_1, E_2 \subset \mathbb{R}^n$.

Proof. We can assume that $\tilde{\Lambda}_0^d(E_1) + \tilde{\Lambda}_0^d(E_2) < \infty$ and the $E_1 \cap E_2 \ne \phi$. Let $\{P_i\}$ be the family $\{Q_{1,k}, Q_{2,k}\}$ where $\{Q_{1,k}\}$, $\{Q_{2,k}\}$ respectively are dyadic covers of E_1 and E_2. Then,

$$E_1 \cup E_2 \subset (\cup_k Q_{1,k})^0 \cup (\cup_k Q_{2,k})^0 \subset (\cup_{j=1}^2 \cup_k Q_{j,k})^0$$

If $\{\tilde{P}_i\} = \{Q_{1,k} \cap Q_{2,k} : Q_{1,k} \cap Q_{2,k} \ne \phi\}$ then $E_1 \cap E_2 \subset (\cup_i \tilde{P}_i)^0$, so

$$\tilde{\Lambda}_0^d(E_1 \cup E_2) + \tilde{\Lambda}_0^d(E_1 \cap E_2) \le \sum_i l(P_i)^d + \sum_i l(\tilde{P}_i)^d \le \sum_k l(Q_{1,k})^d + \sum_j l(Q_{2,j})^d.$$

$\square$

Theorem 3.4. If $\{K_j\}$ is a sequence of compact sets such that $K_j \downarrow K(K = \cap K_j)$, then

$$\lim_{j \to \infty} \tilde{\Lambda}_0^d(K_j) = \tilde{\Lambda}_0^d(K).$$

Proof. It is easy to see the $\tilde{\Lambda}_0^d$ is an outer capacity, i.e., for any $\epsilon > 0$ there exists an open set $G \supset E$ such that $\tilde{\Lambda}_0^d(G) \le \tilde{\Lambda}_0^d(E) + \epsilon$.

Hence with $K = \cap_j K_j$, K_j compact, there is a j_0 such that $K_j \subset G$ for all $j \ge j_0$. And due to the simple fact that $\tilde{\Lambda}_0^d$ is monotone yields

$$\tilde{\Lambda}_0^d(K) \le \tilde{\Lambda}_0^d(K_j) \le \tilde{\Lambda}_0^d(G) \le \tilde{\Lambda}_0^d(K) + \epsilon.$$

$\square$

It is easy to see the $\tilde{\Lambda}_0^d$ is a capacity in the sense of Meyers. What is a bit more interesting is that $\tilde{\Lambda}_0^d$ is in fact a Choquet Capacity. The proof (3.2)(v) can be found in [Y^2], though we will not need it in what follows. They also show that $\tilde{\Lambda}^d$ is not a Choquet Capacity for $0 < d \le n - 1$, a "fact" previously mistakenly claimed in [A5].

3.5   The operator M_α and Hausdorff capacity

In this last section, we give a pseudo potential theoretic description of Hausdorff capacity, in that the operator I_α is replaced by the fractional maximal operator M_α (of Chapters 6, 7, and 9 below).

$$M_\alpha f(x) = \sup_{r>0} r^{\alpha-n} \int_{B(x,r)} |f(y)| \, dy.$$

We begin with:

Theorem 3.5. Let $\Lambda^{\alpha,p}(E) = \inf\{\|f\|_{L^p}^p : f \in L^p_+ \text{ and } M_\alpha f \geq 1 \text{ on } E\}$. Then for K compact

$$\Lambda^{\alpha,p}(K) \sim \Lambda^{n-\alpha p}(K) \quad \alpha p < n, \, p > 1. \tag{3.10}$$

Proof. ($\sim$ means that these are upper and lower bounds with constants independent of the set K.)

It is easy to see that $\Lambda^{\alpha,p}(\cdot)$ is a capacity in the sense of Meyers, and that $M_\alpha \chi_Q(x) \sim |Q|^{\alpha/n}$ for all $x \in Q$. Hence $\Lambda^{\alpha,p} \leq |Q|^{1-\alpha/n}$ for all $x \in Q$. And the first inequality is established by the subadditivity of $\Lambda^{\alpha,p}$.

For the second inequality needed for (3.10): It suffices to prove

$$\Lambda^{n-\alpha p}(M_\alpha f > t) \leq Ct^{-p}\|f\|_{L^p}^p. \tag{3.11}$$

But this follows from the well-known covering lemma to [St1], page 9 (the Vitali lemma). Indeed when $M_\alpha f > t$ there exists a cube Q_x centered at x such that

$$|Q_x|^{\alpha/n-1} \int_{Q_x} |f| > Ct.$$

Hence

$$|Q_x|^{1-\alpha p/n} \leq Ct^{-p} \int_{Q_x} |f|^p,$$

and the result follows from the covering lemma referred to.   □

The reader should compare Theorem 3.5 to the results of Chapter 9, especially to the definition of Riesz-Sobolev capacity $\dot{C}_{\alpha,p}$.

The reason we include Theorem 3.5 in this chapter rather than in Chapter 9 is that it suggests things to come and indicates that Λ^d is really an important tool in HA.

3.6 Notes

3.6.1 Netrusov's capacity $\Lambda^{d;\theta}$ and a Netrusov-Frostman Theorem

In two important papers on the capacities associated with the Besov spaces (of Chapter 14 below), Netrusov introduces a Hausdorff type capacity (content) that he uses to better understand Besov capacity; [N1] and [N2]. He defines

$$\Lambda^{d;\theta}(E) = \inf\left[\sum_{i=1}^{\infty}\left(\sum_{j\in I_i} r_j^d\right)^{\theta}\right]^{1/\theta}$$

for $0 < \theta < \infty$. Here, the infimum is over all countable covers of E by Balls $B(x_j, r_j)$ with $I_i = \{j : 2^{-i-1} \le r_j < 2^{-i}\}$. From [A9], we have the following nice contrast to the Frostman Theorem 3.1.

Theorem 3.6. $\Lambda^{d;\theta}(K) > 0$ iff

(a) $0 < \theta < 1$, there exists a measure μ supported by the compact set K such that

$$\sup_{0<t<1} \int \left[t^{-d}\mu(B(x,t))\right]^{\theta/(1-\theta)} d\mu(x) < \infty; \tag{3.12}$$

(b) $1 < \theta < \infty$, there exists a measure μ supported by K such that

$$\int_0^1 \left(t^{-d}\sup_{x\in K}\mu(B(x,t))\right)^{\theta/(\theta-1)} \frac{dt}{t} < \infty \tag{3.13}$$

And in any case $\mu(K) \sim \Lambda^{d;\theta}(K)$.

3.6.2 *A strong type estimate for M_α*

In analogy to the "capacity strong type" estimates (MaźyaType inequalities) that we shall see in Chapter 10 for Sobolev capacities, we note here a strong type inequality for Hausdorff capacity (see [A4]):

Theorem 3.7. There is a constant $c > 0$ such that

$$\int_0^{\infty} \Lambda^{n-\alpha p}(M_\alpha f \ge t) \, dt^p \le c \, \|f\|_{L^p}^p, \tag{3.14}$$

$\alpha p < n$, $p > 1$; c independent of f.
    Of course (3.11) is the "weak type" estimate here.

Also it has a corollary.

Corollary 3.8. For any Borel measure μ on $\mathbb{R}^n$,

$$\int (M_\alpha f)^p \, d\mu \le c \int |f|^p \, M_{\alpha p}\, \mu \, dx. \tag{3.15}$$

So it might be said that when the capacities taken are Hausdorff Capacities, then the role of the potential operator (as in Chapter 9) is now played by the fractional maximal operator, M_α. Also, we might note that (3.15) is a fractional analogue of the integral inequality of Fefferman-Stein [St2];

$$\int (M_0 f)^p \, d\mu \leq c \int |f|^p M_0 \, \mu \, dx. \tag{3.16}$$

Chapter 4
Choquet Integrals

4.1 Definition and basic properties: sublinear vs. strong subadditivity

One of the new development in the Theory of Morrey spaces - as far as this author is concerned - is the use of Hausdorff capacity and its corresponding Choquet Integral ($\int f \, d\Lambda^d$) in the development of the predual to $L^{p,\lambda}(\mathbb{R}^n)$. This we do in the next chapter once we have thoroughly discussed all the tools that are needed to understand and apply these "non-linear" integrals. And it is because Λ^d is not a measure (i.e., not countable additive on disjoint sets) that we must carefully define and expose all of the relevant properties of such an integral.

We begin with the definition.

Definition For $f \in C_0(\mathbb{R}^n)$ = continuous function on $\mathbb{R}^n$ with compact support, set

$$\int |f|^p \, d\Lambda^d = \int_0^\infty \Lambda^d([|f| \geq t]) \, dt^p. \tag{4.1}$$

In other words, the Choquet Integral of a real valued function is defined distributionally and extended by functional completion. In (4.1) the set $[|f| \geq t]$ is compact and hence the integrals involved are all finite numbers.

But to make use of this functional, we need a metric and in particular a triangle inequality. This was an early key result of Choquet. Our version follows:

Theorem 4.1. Let $C(\cdot)$ be a capacity in the sense of N.G. Meyers which also satisfies (3.2) (vi). Then the Choquet Integral $\int |f| \, dC$ is <u>sublinear</u> iff C is <u>strongly subadditive</u> (i.e., property (3.9) or

$$C(E_1 \cup E_2) + C(E_1 \cap E_2) \leq C(E_1) + C(E_2).) \tag{4.2}$$

© Springer International Publishing Switzerland 2015
D. Adams, *Morrey Spaces*, Applied and Numerical Harmonic Analysis,
DOI 10.1007/978-3-319-26681-7_4

Notice that (4.2) would trivially be true - with equality if C were a linear additive measure. Thus in its simplest form, we intend to show: for $f, g \geq 0$,

$$\int (f + g)\, dC \leq \int f\, dC + \int g\, dC \tag{4.3}$$

iff (4.2) holds for all $E_1, E_2 \subset \mathbb{R}^n$.

Notice that Theorem 4.1 implies that

$$\left(\int |f|^p\, dC \right)^{1/p}, \ 1 \leq p < \infty$$

is a norm by the usual arguments, a norm on what we will eventually call $L^p(C)$. And since C can be taken to be $\tilde{\Lambda}_0^d$, it follows that

$$\left\{ \int |f|^p\, d\Lambda^d \right\}^{1/p}, \ 1 \leq p < \infty, \tag{4.4}$$

defines a quasi-norm (a norm except that the triangle inequality holds with a fixed constant). And it is these facts that will soon send us on our way to the predual promised in the Introduction.

Theorem 4.1 is due to Choquet [C], but we choose to follow the arguments of Topsøe as given in B. Anger [An]. Also, because of the above, we can characterize $L^p(\Lambda^d)$ as consisting of Λ^d - quasi-continuous functions for which (4.4) is finite; i.e., those f such that for any $\epsilon > 0$ there is a subset $E \subset \mathbb{R}^n$ with $\Lambda^d(E) < \epsilon$ and f restricted to $\mathbb{R}^n \backslash E$ is continuous. Here one proceeds as in measure theory.

Proof. Proof of Theorem 4.1. First of all, if C is strongly subadditive and since

$$C(K) = \int X_K\, dC, \tag{4.5}$$

X_K being the characteristic function of the compact set K, it follows that if $K_1 \subset K_2 \subset K_3 \subset \cdots \subset K_n$, then

$$\int \sum_{i=1}^n X_{K_i}\, dC = \sum_{i=1}^n C(K_i). \tag{4.6}$$

The first of these follows from the definition of the Choquet Integral, the second upon breaking up the distribution integral into a sum of integrals from i to $i + 1$, $i = 0, 1, \cdots, n$ and evaluating. This is clearly easy to do here since the sequence $\{K_i\}$ is increasing or at least monotone.

Next observe that (4.4) and (4.5) together with (3.2)(vi) implies

$$\int \inf h_i\, dC = \inf \int h_i\, dC, \tag{4.7}$$

where the infimum is taken over all downward directed sequences $\{h_i\}$ of $h_i \in$ USC$^+$ = non-negative upper semi-continuous functions on $\mathbb{R}^n$. Thus to show (4.3), we may restrict our attention to "compact simple functions," i.e., functions of the form

$$\sum_{i=1}^n \alpha_i X_{K_i}, \ \alpha \in \mathbb{R}^+, \ K_i \text{ compact sets in } \mathbb{R}^n,$$

due to (4.7) and the fact that the sum of two compact simple functions is again a compact simple function. Hence, it suffices to prove that strong subadditivity implies

$$\int \left(\sum_{i=1}^n \alpha_i X_{k_i} \right) dC \le \sum_{i=1}^n \alpha_i C(k_i) \tag{4.8}$$

From (4.6), we have that if $h \in$ USC$^+$

$$\int h \, dC = \inf \left\{ \frac{1}{m} \sum_{i=1}^n C(K_i) : m, n \in \mathbb{N}, K_i \uparrow \text{ and with } \frac{1}{m} \sum_{i=1}^n X_{K_i} \ge h \right\}.$$

And so if we set

$$\hat{C}(h) = \inf \left\{ \sum_{i=1}^n \alpha_i C(K_i) : n \in \mathbb{N}, \ \alpha_i \in \mathbb{R}^+, \ \sum_{i=1}^n \alpha_i X_{K_i} \ge h \right\},$$

then

$$\hat{C}(h) = \inf \left\{ \frac{1}{m} \sum_{i=1}^n C(K_i) : n, m \in \mathbb{N}, \frac{1}{m} \sum_{i=1}^n X_{K_i} \ge h \right\},$$

and hence $\int h \, dC \ge \hat{C}(h)$, for all $h \in$ USC$^+$.

So the final observation will be that strong subadditivity implies $\int h \, dC \le \hat{C}(h)$ and then (4.8). To see this last inequality, we show

$$\sum_{i=1}^n C(K_i) \ge \sum_{i=1}^n C(K_i') \tag{4.9}$$

where $K_i' = \cup \{\cap_{j \in J} K_j : J \subset \{1, 2, \cdots, n\}, |J| = n - i + 1\}$, i.e., ordering the sets into an increasing sequence of compact sets $K_1' \subset, K_2' \subset, \cdots, K_n'$. Equation (4.9) is the key to the proof. When $n = 2$, (4.9) is just strong subadditivity. When $n = 3$, we write

$$C(K_1') + C(K_2') + C(K_3') = C(K_1 \cap K_2 \cap K_3)$$
$$+ C[(K_1 \cap K_2) \cup (K_2 \cap K_3) \cup (K_1 \cap K_3)] + C(K_1 \cup K_2 \cup K_3)$$
$$\leq C(K_1 \cap K_2) + C(F) + C(K_1 \cup K_2 \cup K_3), \qquad (4.10)$$

where $F = (K_2 \cap K_3) \cup (K_1 \cup K_3) = (K_1 \cup K_2) \cap K_3$.

Hence strong subadditivity gives (4.10) as not exceeding

$$C(K_1 \cap K_2) + C(K_1 \cup K_2) + C(K_3) \leq C(K_1) + C(K_2) + C(K_3).$$

This then shows how to go from two sets to three sets in the induction process. The rest is similar but messy.

For the converse, we can simply write

$$\int X_{K_1 \cup K_2} \, dC + \int X_{K_1 \cap K_2} \, dC = \int (X_{K_1 \cup K_2} + X_{K_1 \cap K_2}) \, dC$$
$$= \int (X_{K_1} + X_{K_2}) \, dC \leq \int X_{K_1} \, dC + \int X_{K_2} \, dC.$$

$\square$

4.2 Adams-Orobitg-Verdera Theorem

With these preliminaries out of the way, we are now ready for the real substance - the boundedness of the Hardy-Littlewood maximal function on the spaces $L^p(\Lambda^d)$. The following theorem is due to Adams [A5] in the case p=1 and to Orobitg-Verdera for $p < 1$. For $p > 1$, the result is of course classical. The case $p = 1$ was accomplished using H^1-BMO duality (see [St2] or [To]), whereas $p < 1$ was done the old fashioned way, via a covering lemma. For our purposes, we need only a small part of that proved in [OV].

Here, of course

$$M_0 f(x) = \sup_{r>0} \fint_{B(x,r)} |f(y)| \, dy.$$

Theorem 4.2. There is a constant A depending only on d, n, and p such that

$$\int (M_0 f)^p \, d\Lambda^d \leq A \cdot \int |f|^p \, d\Lambda^d \qquad (4.11)$$

for $0 < d \leq n$, $d/n < p < \infty$.

Proof: We begin with a lemma:

Lemma 4.3. If X_Q is the characteristic function of a cube Q (sides parallel to the axes), then

$$\int M_0(X_Q)^p \, d\Lambda^d \leq C \cdot l(Q)^d \quad \text{for } d/n < p.$$

Proof. Let x_0 be the center of Q, then

$$M_0(X_Q)(x) \leq C \cdot \inf\left(1, \frac{l(Q)^n}{|x - x_0|^n}\right),$$

for $x \in \mathbb{R}^n$. Then

$$\int M_0(X_Q)^p \, d\Lambda^d \leq C \cdot l(Q)^d + C \int_0^1 l(Q)^d t^{-d/np} \, dt$$

$$= C' \cdot l(Q)^d,$$

since $d/np < 1$.

With this, we proceed as follows: for $f \geq 0$ and

$$\{x : 2^k < f(x) \leq 2^{k+1}\} \subset \cup_j Q_j^{(k)}$$

for some non-overlapping dyadic cubes $Q_j^{(k)}$, then

$$\sum_j l(Q_j^{(k)})^d \leq 2 \, \tilde{\Lambda}^d \left(\{x : 2^k < f(x) \leq 2^{k+1}\}\right)$$

Setting $g = \sum_k 2^{(k+1)p} X_{A_k}$, $A_k = \cup_j Q_j^{(k)}$, we have

$$f^p \leq g.$$

So now if $d/n < p < 1$, then

$$f \leq \sum_k 2^{k+1} X_{A_k},$$

and

$$(M_0 f)^p \leq \sum_k 2^{(k+1)p} \cdot \sum_j M_0 \left(X_{Q_j^{(k)}}\right)^p$$

because $p < 1$. Consequently,

$$\int (M_0 f)^p \, d\Lambda^d \le c \cdot \sum_k 2^{(k+1)p} \cdot \sum_j l\left(Q_j^{(k)}\right)^d$$

$$\le c \int f^p \, d\Lambda^d.$$

□

For the corresponding "weak-type" estimate at $p = d/n$, the covering lemma mentioned earlier comes into play. See [OV] and below in Notes.

4.3   Notes

4.3.1   *Further estimates for $M_\alpha f$*

Using a covering lemma in [OV], the authors were able to prove the following much more difficult result.

Theorem 4.4. For $0 < d \le n$, $p = d/n$,

$$\Lambda^d\left\{[M_0 f > t]\right\} \le C \, t^{-d/n} \cdot \int |f|^{d/n} \, d\Lambda^d. \tag{4.12}$$

But then the present author responded by proving the following extension of Theorems 4.2 and 4.4:

Theorem 4.5. For $0 < d \le n$, $0 < \alpha < n$,

(a) $p \le q$

  (i) $\dfrac{d}{n} < p < \dfrac{d}{\alpha}$ and $\delta = q(d - \alpha p)/p$, then

$$||M_\alpha f||_{L^{q,p}(\Lambda^\delta)} \le A_1 ||f||_{L^p(\Lambda^d)}; \tag{4.13}$$

  (ii) $p = \dfrac{d}{n}$

$$||M_\alpha f||_{L^{q,\infty}(\Lambda^\delta)} \le A_2 ||f||_{L^{d/n}(\Lambda^d)} \tag{4.14}$$

  with $q = \dfrac{\delta}{(n - \alpha)}$, $\delta \ge \dfrac{d}{n}(n - \alpha)$;

  (iii) $p = \dfrac{d}{\alpha}$

$$||M_\alpha f||_{L^\infty} \le A_3 ||f||_{L^{d/\alpha}(\Lambda^d)}. \tag{4.15}$$

Here $L^{q,p}(\Lambda^d)$ is the corresponding Lorentz space with respect to Λ^d (see [St2]).

(b) $q < p$

$$||M_\alpha f||_{L^q(\omega\Lambda^{d-\alpha p})} \leq A_4 ||f||_{L^p(\Lambda^d)}, \quad \alpha p < d, \quad (4.16)$$

iff

$$||\omega||_{L^{p/(p-q)}(\Lambda^{d-\alpha p})} < \infty.$$

The reader is refereed to [A7] for more details.

4.3.2 Speculations on weighted Hausdorff Capacity

In [A6], the author asked the question:

Is weighted capacity related to the Choquet Integral
of the weight with respect to the unweighted capacity?

We may have more to say on this point in Chapter 9 when various L^p- capacities are examined. But for now we might just ask this question for the weighted Hausdorff capacities of [Ni], i.e., suppose ω is an A_∞- weight in the sense of Muckenhoupt [St2] and set

$$\Lambda^{d/\omega}(E) = \inf \left\{ \sum_j r_j^d \cdot \fint_{B(x_j, r_j)} \omega(y)\, dy : E \subset \cup_j B(x_j, r_j) \right\}. \quad (4.17)$$

One can define the usual dyadic version of this weighted set function, so the question becomes: do such capacities satisfy all the properties discussed in Chapter 3? Perhaps of more interest is: when is the following true?

$$\int f^p \, d\Lambda^{d/\omega} \sim \int f^p \, \omega \, d\Lambda^d, \quad (4.18)$$

$f \geq 0$? And what about an analogue of Theorem 4.5 for the fractional maximal operator M_α, with Λ^d replaced by $\Lambda^{d/\omega}$? One can think of (4.18) as a sort of Radon-Nikodym Theorem for weighted Hausdorff Capacities. More on this on Chapter 9.

Chapter 5
Duality for Morrey Spaces

5.1 Dual of $L^1(\Lambda^d)$

A precursor of our duality theory for the spaces $L^{p,\lambda}(\mathbb{R}^n)$ is

Theorem 5.1. The dual of $L^1(\Lambda^d)$ is the Morrey space of signed Borel measures $\mu \in L^{1,d}$.

Proof. : The norm of $\mu \in L^{1,d}$ is

$$\|\mu\|_{L^{1,d}} = \sup_{x,r>0} r^{-d} |\mu|(B(x,r)) < \infty,$$

where $|\mu|$ = total variation measure = $\mu^+ - \mu^-$.
  The duality is given by

$$l(\varphi) = \int \varphi \, d\mu, \qquad \varphi \in C_0. \tag{5.1}$$

In fact, covering E by $\{B(x_j, r_j)\}$, we have

$$|\mu|(E) \leq \sum_j |\mu|(B(x_j, r_j))$$

$$\leq \|\mu\|_{L^{1,d}} \cdot \sum_j r_j^d.$$

Hence it follows that

$$|\mu|(E) \leq \|\mu\|_{L^{1,d}} \Lambda^d(E).$$

© Springer International Publishing Switzerland 2015
D. Adams, *Morrey Spaces*, Applied and Numerical Harmonic Analysis,
DOI 10.1007/978-3-319-26681-7_5

And then

$$|\hat{l}(\varphi)| \le \int |\varphi| d|\mu| = \int_0^\infty |\mu|([|\varphi| > t]) \, dt$$

$$\le ||\mu||_{L^{1,d}} \int_0^\infty \Lambda^d([|\varphi| > t]) \, dt$$

$$= ||\mu||_{L^{1,d}} ||\varphi||_{L^1(\Lambda^d)}.$$

For the converse, since $C_0 \subset L^1(\Lambda^d)$, l will be given (5.1) for some μ. But then for $\psi \in C_0(\mathbb{R}^n)$

$$\left| \int \psi \, d|\mu| \right| \le \sup \left\{ \int \varphi \, d\mu : \varphi \in C_0 \,\&\, |\varphi| \le |\psi| \right\}$$

$$\le \sup \left\{ ||l|| \int |\varphi| \, d\Lambda^d : \varphi \in C_0 \,\&\, |\varphi| \le |\psi| \right\}$$

$$\le ||l|| \int |\psi| \, d\Lambda^d.$$

Thus if $\psi = 1$ on $B(x, r)$ and $\psi = 0$ on $B(x, r + \epsilon)^C$, then

$$|\mu|(B(x, r)) \le ||l|| \, \Lambda^d \, (B(x, r + \epsilon)) = ||l|| \cdot (r + \epsilon)^d$$

for any $\epsilon > 0$. So

$$||\mu||_{L^{1,d}} \le ||l||.$$
□

Theorem 5.1 first appeared in [A5] in 1988. We now use it to advance to the duality for $L^{p,\lambda}$. The following is from [AX2], 2004.

Theorem 5.2. Let $1 < p < \infty$, $0 < \lambda < n$, then

$$||f||_{L^{p,\lambda}} = \sup_\omega \left(\int_{\mathbb{R}^n} |f|^p \, \omega \, dx \right)^{1/p} \tag{5.2}$$

where the supremum is taken over all non-negative functions ω on $\mathbb{R}^n$ such that

$$||\omega||_{L^1(\Lambda^d)} \le 1, \quad d = n - \lambda. \tag{5.3}$$

Proof. : From Theorem 5.1,

$$\int_{\mathbb{R}^n} |f|^p \, \omega \, dx \le ||\omega||_{L^1(\Lambda^d)} ||f||_{L^{p,\lambda}}^p \le ||f||_{L^{p,\lambda}}^p.$$

On the other hand, if $\omega_0 = X_{B(x_0,r_0)} \cdot r_0^{\lambda-n}$, then

$$||\omega_0||_{L^1(\Lambda^d)} = \int_{B(x_0,r_0)} r_0^{\lambda-n} \, d\Lambda^d = 1.$$

Thus, ω_0 satisfies (5.3) and

$$||f||_{L^{p,\lambda}} = \sup_{x_0,r_0>0} \left(\int_{\mathbb{R}^n} |f|^p \, r_0^{\lambda-n} \, X_{B(x_0,r_0)} \, dx \right)^{1/p}$$

$$= \sup_{x_0,r_0>0} \left(\int |f|^p \, \omega_0 \right)^{1/p} \leq \sup_{\omega} \left(\int |f|^p \, \omega \, dx \right)^{1/p}.$$

$\square$

5.2   Three equivalent predual spaces $X^{p,\lambda}$, $K^{p,\lambda}$, $Z^{p,\lambda}$

From this, we can now give three spaces that describe the predual of $L^{p,\lambda}(\mathbb{R}^n)$. We begin with $X^{p,\lambda}$, from [AX2], though given a different name here since we wish to reserve the letter H for our final version (see Theorem 5.5).

Definition $g \in X^{p,\lambda}$, $1 < p < \infty$, $0 < \lambda < n$, if

$$||g||_{X^{p,\lambda}} = \inf_{\omega} \left(\int_{\mathbb{R}^n} |g|^p \, \omega^{1-p} \, dx \right)^{1/p} < \infty, \tag{5.4}$$

where the infimum is over all non-negative functions ω on $\mathbb{R}^n$ satisfying (5.3).

Theorem 5.3. For $1 < p < \infty$, $0 < \lambda < n$, $p' = p/(p-1)$, the predual to $L^{p,\lambda}$ is $X^{p',\lambda}$ under the following

$$<f, g> = \int_{\mathbb{R}^n} f(x) \cdot g(x) \, dx.$$

Moreover,

$$||f||_{L^{p,\lambda}} = \sup_{g} \left| \int f \cdot g \, dx \right|$$

where $f \in L^{p,\lambda} = L^{p,\lambda}(\mathbb{R}^n)$ and the supremum is over all functions $g \in X^{p',\lambda}$ with $||g||_{X^{p',\lambda}} \leq 1$.

Proof. : Let $f \in L^{p,\lambda}$ and $g \in X^{p',\lambda}$, then

$$| <f\, g> | \leq \int_{\mathbb{R}^n} |f|\, |g|\, \omega^{1/p}\, \omega^{-1/p}\, dx$$

$$\leq \left(\int_{\mathbb{R}^n} |f|^p\, \omega\, dx \right)^{1/p} \left(\int |g|^{p'}\, \omega^{1-p'}\, dx \right)^{1/p'}.$$

Conversely, if L is a bounded linear functional on $X^{p',\lambda}$, with norm $||L||$, it induces a bounded linear functional on $L^{p'}(B(x_0, r_0))$ corresponding to some function $f^B \in L^p(B(x_0, r_0))$. Taking $f^{B_j} = f^{B_{j+1}}$, $B_j = B(0,j)$ $j = 1,2,3,\cdots$, we get a single function of f on $\mathbb{R}^n$ that is locally in L^p such that

$$L(g) = \int_{\mathbb{R}^n} f \cdot g\, dx$$

with $g \in X^{p',\lambda}$ and supported on some ball of $\mathbb{R}^n$. So take $g = X_{B(x_0, r_0)} \cdot |f|^p \cdot f^{-1}$ and then

$$\int_{B(x_0, r_0)} |f|^p\, dx = L(g) \leq ||L||\, ||g||_{X^{p',\lambda}}$$

$$= ||L|| \left(r_0^{(\lambda-n)(1-p')} \int_{B(x_0, r_0)} |f|^{(p-1)p'}\, dx \right)^{1/p'},$$

upon setting $\omega_0 = X_{B(x_0, r_0)} \cdot r_0^{\lambda-n}$ and then noting

$$||g||_{X^{p',\lambda}} \leq \left(r_0^{(\lambda-n)(1-p')} \int_{B(x_0, r_0)} |g|^{p'}\, dx \right)^{1/p'}.$$

Hence $||f||_{L^{p,\lambda}} \leq ||L||$. $\qquad\qquad\qquad\qquad\qquad\qquad\qquad\qquad\qquad\qquad\square$

The first predual space formulated for any Morrey type space is clearly that formulated for $\mathscr{L}^{p,\lambda}$, $0 < \lambda < n$ in [Z] in 1986. Her predual space, denoted below by $Z_0^{p',\lambda}$, is given via an atomic decomposition and is discussed fully below in Section 5.4. What we do next is give an atomic version of the predual of $L^{p,\lambda}$. We continue to use Z notation in her honor, though one should note the small but important difference between $Z^{p,\lambda}$ and $Z_0^{p,\lambda}$. For this, we shall say that the function $a(x)$ is a $(p', n-\lambda)$-atom if

(i) supp $a \subset B$ = ball of $\mathbb{R}^n$
(ii) $||a||_{L^{p'}} \leq |B|^{-(n-\lambda)/np}$.

And then we define $Z^{p',\lambda}$ as those f on $\mathbb{R}^n$ such that

$$||f||_{Z^{p',\lambda}} = \inf \left\{ ||\{c_k\}||_{l^1} : f = \sum_{k \geq 0} c_k\, a_k \right\} < \infty, \qquad (5.5)$$

where the infimum is over all such decompositions of f into sums of constant multiples of $(p', n - \lambda)$-atoms.

A second version of a predual to $L^{p,\lambda}$ seems to have been given by Kalita [K] in 1998. His formulation came out of his work on partial differential equations, and it is clear that the construction of $X^{p',\lambda}$ above has been modeled on Kalita's, though the weight functions are different. We shall say that $f \in K^{p',\lambda}$ if

$$\|f\|_{K^{p',\lambda}} = \inf_{\sigma} \left(\int |f|^{p'} \omega_{\sigma}^{1-p'} \, dx \right)^{1/p'} < \infty, \tag{5.6}$$

where

$$\omega_{\sigma}(x) = \int_{\mathbb{R}_+^{n+1}} r^{-(n-\lambda)} X_{\mathbb{R}_+}(r - |x - y|) \, d\sigma(y, r)$$

and the infimum is over all measures σ on the upper half space $\mathbb{R}_+^{n+1} = \{(x, r) : x \in \mathbb{R}^n, r > 0\}$ normalized with

$$\sigma(\mathbb{R}_+^{n+1}) = 1. \tag{5.7}$$

So we now have

Theorem 5.4. For $1 < p < \infty$, $p' = p/(p-1)$ and $0 < \lambda < n$,

$$X^{p',\lambda} = Z^{p',\lambda} = K^{p',\lambda} \tag{5.8}$$

with equivalent norms.

Proof. : The proof (5.8) appears in [AX2], 2004. But because we have no need of this below we defer the reader to the literature. We do need $X^{p',\lambda}$, however, since it leads into our desired formulation of the predual to $L^{p,\lambda}$ that seems to be closest to the spirit of classical HA as given in the classics [St2] and [To]. □

5.3 The predual $H^{p',\lambda}$

Finally, we have

Theorem 5.5. $X^{p,\lambda}$ is equal to

$$H^{p,\lambda} = \left\{ f \in L_{loc}^p : \|f\|_{H^{p,\lambda}} = \inf_{\omega} \left(\int |f(y)|^p \, \omega(y)^{1-p} \, dy \right)^{1/p} < \infty \right\} \tag{5.9}$$

where the infimum is over all weights $\omega \in A_1$ such that

$$\int_{\mathbb{R}^n} \omega \, d \, \Lambda^{n-\lambda} = \int_0^\infty \Lambda^{n-\lambda}([\omega > t]) \, dt \leq 1, \qquad (5.10)$$

with equivalent norms; $1 < p < \infty$, $0 < \lambda < n$.

Recall that a non-negative locally integrable function $\omega \in A_1$ (Muckenhoupt's weight scale) iff

$$\fint_Q \omega \, dx \leq c_1 \cdot \inf_Q \omega$$

for some constant c_1. Furthermore, $\omega \in A_p$ iff

$$\left(\fint_Q \omega \, dx \right) \left(\fint_Q \omega^{-\frac{1}{p-1}} \, dx \right)^{p-1} \leq c_p < \infty,$$

for all cubes Q (sides parallel to axes). And note that

$$\omega \in A_1 \implies \omega \in A_p, \text{ for all } p > 1.$$

These weights will have an important role to play later in Chapter 8. For a full discussion of the theory of A_p weights, the reader is referred to [To].

Proof. Clearly $\|f\|_{H^{p,\lambda}} \geq \|f\|_{X^{p,\lambda}}$. For the reverse inequality, we let ω be a weight in the construction of $X^{p,\lambda}$ and modify it to:

$$\omega_\theta = (M_0 \, \omega^{1/\theta})^\theta, \ 0 < \theta < 1.$$

Theorem 4.2 gives

$$\int \omega_\theta \, d\Lambda^{n-\lambda} \leq c_0 \int \omega \, d\Lambda^{n-\lambda} \leq c_0$$

provided $(n - \lambda)/n < \theta \leq 1$, $0 < \lambda < n$.

On the other hand, we know from the classical constructions of Coifman-Rochberg (see again [To] or [St2]) that ω_θ/c_0 belongs to A_1 and satisfies (5.3). Hence

$$\|f\|_{H^{p,\lambda}}^p \leq \int |f(y)|^p \omega_\theta(y)^{1-p} \, dy \leq c \int |f(y)|^p \omega(y)^{1-p} \, dy,$$

giving

$$\|f\|_{H^{p,\lambda}} \leq c \, \|f\|_{X^{p,\lambda}}.$$

$\square$

Thus in the language of Functional Analysis,

$$(H^{p',\lambda})^* = L^{p,\lambda}.$$

Notice, that the Zorko space $VL^{p,\lambda}$ plays a similar role to VMO, the Sarason functions of vanishing mean oscillation, in the Fefferman-Stein duality for our duality; i.e.

$$VL^{p,\lambda} \xrightarrow{\quad*\quad} H^{p',\lambda} \xrightarrow{\quad*\quad} L^{p,\lambda}$$

as in

$$VMO \xrightarrow{\quad*\quad} H^1 \xrightarrow{\quad*\quad} BMO.$$

5.4   The space $Z_0^{p,\lambda}$ and $Z^{p,\lambda}$

Finally in this last section on duality, we look briefly at the exact preduality results proved by Zorko in [Z], for Theorem 5.4 involving $Z^{p',\lambda}$ is not her result. To be precise, we first need $(p', n-\lambda)$ - atoms with mean value zero, i.e. $Z_0^{p',\lambda}$ will denote those f such that

$$\|f\|_{Z_0^{p',\lambda}} = \inf \int \left\{ \|\{c_k\}\|_{l^1} : f = \sum_{k\geq 0} c_k \, a_k \right\} < \infty$$

where the infimum is now over all such decompositions where a_k is a $(p', n-\lambda)$-atom such that

$$\int_{\mathbb{R}^n} a_k \, dx = 0.$$

The result of Zorko is now:

Theorem 5.6. $(Z_0^{p',\lambda})^* = \mathcal{L}^{p,\lambda}$, $1 < p < \infty$, $0 < \lambda < n$

Recall that only $\mathcal{L}^{p,\lambda}$ is the same as the Morrey space $L^{p,\lambda}$. This result is more in line with the duality results of Fefferman-Stein, but we mention it because it will be useful when we extend the Morrey Theory from $\mathbb{R}^n$ to complete Riemannian manifolds (non-compact case) in Chapter 17. We leave the details of the proof of Theorem 5.6 to the interested reader- for as one easily can see, there are a multitude of open problems regarding Morrey theory on such manifolds.

5.5   Notes

(1) By analogy, one would expect a space $V\mathscr{L}^{p,\lambda}$, $1 < p < \infty$, $0 < \lambda < n$, and the corresponding result.

$$V\mathscr{L}^{p,\lambda} \overset{*}{\longrightarrow} Z_0^{p',\lambda} \overset{*}{\longrightarrow} \mathscr{L}^{p,\lambda}$$

where we recall from Chapter 2 that $\mathscr{L}^{p,\lambda}$ is the space of functions in the Campanato class

$$\sup_Q |Q|^{\lambda/n} \fint_Q |f(x) - f_Q|^p \, dx \;=\; \|f\|^p_{\mathscr{L}^{p,\lambda}}$$

f_Q = mean of f over the cube Q. And $\mathscr{L}^{p,\lambda} \supsetneq L^{p,\lambda}$ - with equivalence, locally (see (2.1)); $0 < \lambda < n$, $1 < p < \infty$. Recall $\mathscr{L}^{p,\lambda} = \mathscr{L}^{p,\lambda}(\mathbb{R}^n)$. Furthermore, what distinguishes $\mathscr{L}^{p,\lambda}$ from $L^{p,\lambda}$ at ∞?

(2) Do the same for $\mathscr{L}_k^{p,\lambda}$, $k \geq 1$, k = degree of polynomial approximates

$$\sup_Q |Q|^{\lambda/n} \fint_Q |f(x) - P_Q|^p \, dx \;=\; \|f\|^p_{\mathscr{L}_k^{p,\lambda}}$$

with $\deg P_Q \leq k$.

The reader might find the paper [GJ], 1978, of interest here in distinguishing $\mathscr{L}^{p,\lambda}$ from $L^{p,\lambda}$.

(3) Diagram of duality

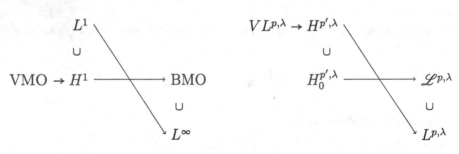

L^1		$VL^{p,\lambda} \to H^{p',\lambda}$
U		U
VMO $\to H^1$ ⟶ BMO		$H_0^{p',\lambda}$ ⟶ $\mathscr{L}^{p,\lambda}$
U		U
L^∞		$L^{p,\lambda}$

Fefferman- Stein   Adams - Xiao
(1972)   (2012)

Chapter 6
Maximal Operators and Morrey Spaces

6.1 M_0 on $L^{p,\lambda}$ - two proofs

We have already introduced the maximal operators $M_\alpha f$, $0 < \alpha < n$, in Section 3.3 and in Chapter 4. Recall

$$M_\alpha f(x) = \sup_{r>0} r^\alpha \fint_{B(x,r)} |f(y)| dy.$$

Our purpose here is to describe several results involving these operators that will be important later. But we begin with the simple assertion that M_0, the classical Hardy-Littlewood maximal function, is continuous from $L^{p,\lambda}$ to $L^{p,\lambda}$, $1 < p < \infty$, $0 < \lambda \le n$. Of course the case $\lambda = n$ is the famous result of Hardy-Littlewood-(Weiner). We give two proofs of this: the first based on the method of [A3], and the second based on our duality of Chapter 5.

Theorem 6.1. For any $f \in L^{p,\lambda}$, $1 < p < \infty$, $0 < \lambda \le n$, there is a constant independent of f such that

$$||M_0 f||_{L^{p,\lambda}} \le c\, ||f||_{L^{p,\lambda}}. \tag{6.1}$$

Proof. : (1) We write $f = f_r + f^r$, where

$$f_r(x) = \begin{cases} f(x), & \text{for } |x - x_0| < 2r \\ 0, & \text{otherwise.} \end{cases}$$

Then

$$\int_{|x-x_0|<r} (M_0 f)^p\, dx \le 2^p \int_{|x-x_0|<r} (M_0 f_r)^p\, dx + 2^p \int_{|x-x_0|<r} (M_0 f^r)^p\, dx$$

$$= 2^p (I_1 + I_2).$$

© Springer International Publishing Switzerland 2015
D. Adams, *Morrey Spaces*, Applied and Numerical Harmonic Analysis,
DOI 10.1007/978-3-319-26681-7_6

$$I_1 \leq c \int_{\mathbb{R}^n} |f_r(y)|^p \, dy = C \int_{|y-x_0|<2r} |f(y)|^p \, dy \leq C \, r^{n-\lambda} \, \|f\|_{L^{p,\lambda}}^p,$$

by the classical result $M_0 : L^p \to L^p$; see [St1]. For I_2, we note

$$\frac{1}{\rho^n} \int_{|y-x|<\rho} |f^r(y)| \, dy = \frac{1}{\rho^n} \int_{(|x-y|<\rho) \cap (|x_0-y|>2r)} |f(y)| \, dy \qquad (6.2)$$

$$\leq c \left(\rho^{\lambda-n} \int_{|x-y|<\rho} |f(y)|^p \, dy \right)^{1/p} \cdot \rho^{-\lambda/p}$$

with also $|x - x_0| < r$. Hence (6.2) does not exceed

$$c \, \|f\|_{L^{p,\lambda}} \, r^{-\lambda/p}$$

which gives $I_2 \leq c \cdot r^{n-\lambda} \, \|f\|_{L^{p,\lambda}}^p$ and the result follows.   □

Theorem 6.1 was claimed by Chiarenza-Frasca in [CF] in 1987.

Proof. (2). Here we use Theorem 5.2. Choosing the weight $\omega \in A_1$, it follows from the weighted norm inequalities of Muckenhoupt (see [St2] or [To]) that

$$\int (M_0 f)^p \, \omega \, dx \leq c \int |f(y)|^p \, \omega(y) \, dy. \qquad (6.3)$$

And finally notice that $\omega(x) = r^{\lambda-n} \chi_{B(x_0,r)}(x)$ is an A_1-weight. This gives the result.   □

Notice, by the same argument, we have

Theorem 6.2. For any $g \in H^{p,\lambda}$, $1 < p < \infty$, $0 < \lambda \leq n$,

$$\|M_0 \, g\|_{H^{p,\lambda}} \leq c \, \|g\|_{H^{p,\lambda}} \qquad (6.4)$$

again with c independent of g.

Proof. We now take the weight ω^{1-p}, where $\omega \in A_1$, and hence $\omega^{1-p} \in A_p$.   □

We can now make the following duality argument (using the Fefferman-Stein maximal inequality (3.16))

$$\left| \int M_0 f \cdot g \, dx \right| \leq c \int |f| \cdot M_0 \, g \leq c \, \|f\|_{L^{p,\lambda}} \cdot \|M_0 \, g\|_{H^{p',\lambda}}$$

$$\leq c \, \|f\|_{L^{p,\lambda}} \, \|g\|_{H^{p',\lambda}}.$$

So

$$||M_0 f||_{L^{p,\lambda}} = \sup_{||g||_{H^{p',\lambda}} \leq 1} \left| \int M_0 f \cdot g \, dx \right| \leq c \, ||f||_{L^{p,\lambda}}.$$

6.2 $||I_\alpha \mu||_{L^{p,\lambda}} \sim ||M_\alpha \mu||_{L^{p,\lambda}}$, $||I_\alpha \mu||_{H^{p,\lambda}} \sim ||M_\alpha \mu||_{H^{p,\lambda}}$

Our next result will be very useful with regard to estimating Wolff potentials and consequently the singular sets of Morrey potentials in Chapter 10.

Theorem 6.3. For any Borel measure μ,

(i) $||I_\alpha\mu||_{L^{p,\lambda}} \sim ||M_\alpha\mu||_{L^{p,\lambda}}$
    and

(ii) $||I_\alpha\mu||_{H^{p,\lambda}} \sim ||M_\alpha\mu||_{H^{p,\lambda}}$

for $1 < p < \infty$, and $0 < \lambda < n$.

Proof. Notice that we always have

$$M_\alpha\mu(x) \leq c \, I_\alpha \, \mu(x)$$

for all $x \in \mathbb{R}^n$ and any Borel measure μ on $\mathbb{R}^n$. Hence we only need to estimate $||I_\alpha\mu||$ in $L^{p,\lambda}$ and $H^{p,\lambda}$ in terms of $||M_\alpha\mu||$ in the same respective spaces. To get these results, we will rely on a fundamental duality of Fefferman-Stein suitably modified (see [St2] pages 146–149):

- If $f \in L^{p_1 + p_2}(\mathbb{R}^n)$, $1 < p_1 < p_2 < \infty$, (i.e., f can be written as a sum $f = f_1 + f_2$ for $f_1 \in L^{p_1}$ and $f_2 \in L^{p_2}$) and $g \in C_0^\infty(\mathbb{R}^n)$, then there is a constant $C > 0$ and independent of f and g such that

$$\left| \int_{\mathbb{R}^n} f(x) \, g(x) \, dx \right| \leq C \int_{\mathbb{R}^n} f^\#(x) \cdot M_0 \, g(x) \, dx \qquad (6.5)$$

Here $f^\#(x)$ is the Fefferman-Stein Sharp Function:

$$f^\# = \sup_Q \fint_Q |f(y) - f_Q| \, dy,$$

where the supremum is over all cubes (sides parallel to axes) centered at x. We intend to apply (6.5) to the case $f(x) = I_\alpha\mu$, initially with μ having compact support. Also, we will need the following estimate

$$(I_\alpha\mu)^\#(x) \sim M_\alpha\mu(x), \quad x \in \mathbb{R}^n. \qquad (6.6)$$

We will outline the proof of (6.6) below in Section 6.3, it first appeared in [A3] in 1975.

So we proceed as follows

$$\left| \int I_\alpha \mu \cdot g \, dx \right| \leq c \int (I_\alpha \mu)^{\#} \cdot M_0 \, g \, dx$$

$$\leq c \int M_\alpha \mu \cdot M_0 g \, dx \; \leq c \; ||M_\alpha \mu||_{L^{p,\lambda}} \cdot ||M_0 g||_{H^{p',\lambda}}$$

$$\leq c \; ||M_\alpha \mu||_{L^{p,\lambda}} \cdot ||g||_{H^{p',\lambda}}.$$

With this, we can pass to more general functions g via the monotone convergence theorem or suitable limit theorem - then apply $(H^{p',\lambda}, L^{p,\lambda})$ duality. $\square$

A different "proof" of this result was given initially in [AX2], which turned out to have a deficiency - which was eliminated later in [AX3].

For the proof of (ii) in Theorem 6.3, we proceed similarly though using the $(VL^{p,\lambda}, H^{p',\lambda})$ duality instead.

6.3   Proof of (6.6)

(1) We outline the proof for (6.6) as given in [A3]

$$(i) \quad (I_\alpha \mu)^{\#}(x) \leq c \cdot M_\alpha \mu(x). \tag{6.7}$$

Here we set $\mu = \mu_r + \mu^r$ as in Theorem 6.1, and write

$$\int_{|x-x_0|<r} \int_{|y-x_0|<2r} |y-x|^{\alpha-n} \, d\mu(y) \, dx$$

$$\leq \int_{|y-x_0|<2r} \int_{|y-x|<3r} |y-x|^{\alpha-n} \, dx \, d\mu(y)$$

$$\leq cr^\alpha \int_{|y-x_0|<2r} d\mu(y) \leq C \, r^n \, M_\alpha \mu(x_0).$$

Hence

$$\int_{|x-x_0|<r} |I_\alpha \mu_r(x) - (I_\alpha \mu_r)_r(x_0)| \, dx \leq C \, r^n \, M_\alpha \mu(x).$$

Then applying the mean value theorem to the difference

$$|x-z|^{\alpha-n} - |y-z|^{\alpha-n}, \; \text{with} \; \begin{array}{l} |x-x_0| < r \\ |y-x_0| < r \end{array} \; \text{and} \; |z-x_0| > 2r,$$

we have

$$|I_\alpha \mu^r(x) - I_\alpha \mu^r(x_0)| \leq c\, r \int_{|z-x_0|>2r} |z - x_0|^{\alpha-n-1}\, d\mu(z)$$

$$= c\, r \sum_{k=1}^{\infty} \int_{2^k r < |z-x_0| < 2^{k+1}\, r} |z - x_0|^{\alpha-n-1}\, d\mu(z) \leq c \cdot M_\alpha \mu(x_0).$$

$$(ii) \quad M_\alpha \mu(x) \leq c \cdot (I_\alpha \mu)^{\#}(x). \tag{6.8}$$

Here we need two lemmas.

Lemma 6.4. If $\phi \in C_0^\infty(\mathbb{R}^n)$ and f such that

$$\int \left| I_\alpha f(x) \right| (1 + |x|)^{-n-\alpha}\, dx < \infty,$$

then

$$\int \phi(x) f(x)\, dx = c \int I_\alpha f(x) \cdot I_\beta (\Delta^k \phi)(x)\, dx \tag{6.9}$$

with c independent of f and ϕ.

Here Δ^k is the k-th power of the Laplacian $k \in \mathbb{N}$, and $\alpha + \beta = 2k$, $0 < \beta < n$. This Lemma then follows via Fourier Transform analysis; see [W] for details. And from Fefferman-Stein [St2], we have

Lemma 6.5. If $F \in L^1_{loc}$, then

$$\int \frac{|F(x) - F_1(0)|}{(|x|^{n+\alpha} + 1)} \leq c \cdot F^{\#}(0). \tag{6.10}$$

Proof. Of (6.8): Let $F_1(0)$ be the average of F on the ball $B(0,1)$. Then we need only prove

$$\int_{|x|<1} f(x)\, dx \leq c \cdot (I_\alpha f)^{\#}(0).$$

with $\phi \in C_0^\infty(\mathbb{R}^n)$, supp $\phi \subset B(0,2)$ and $\phi = 1$ in $B(0,1)$ we can write

$$|I_\beta(\Delta^k \phi)(x)| \leq c \cdot (1 + |x|)^{-n-\alpha}.$$

So with this on the left side of (6.9), we get

$$\left| \int I_\alpha f \cdot I_\beta(\Delta^k \phi)\, dx \right| \leq C \int |I_\alpha f - (I_\alpha f)_1(0)| (1 + |x|)^{-n-\alpha}\, dx$$

which does not exceed $(I_\alpha f)^\#(0)$. And

$$\int_{|x|>1} \frac{|I_\alpha f(x) - (I_\alpha f)_1(0)|}{|x|^{n+\alpha}}\, dx \leq C\,(I_\alpha f)^\#(0),$$

by Lemma 6.5.

Of course $f\, dx$ can then be replaced throughout by $d\mu$.   □

6.4 Notes

Using the duality $(Z_0^{p,\lambda}, \mathscr{L}^{p,\lambda})$, as mentioned in 4.5 (1), one should be able to get the analogue of Theorem 6.3 (i) and (ii) as

$$||I_\alpha \mu||_{\mathscr{L}^{p,\lambda}} \sim ||M_\alpha \mu||_{\mathscr{L}^{p,\lambda}}$$

and

$$||I_\alpha \mu||_{Z_0^{p,\lambda}} \sim ||M_\alpha \mu||_{Z_0^{p,\lambda}}.$$

These might have some consequences for estimating corresponding Wolff potentials associated with these Campanato-and predual- spaces. See Chapter 7.

Chapter 7
Potential Operators on Morrey Spaces

7.1 $I_\alpha : L^{p,\lambda} \longrightarrow L^{\tilde{p},\lambda} \cap L^{p,\lambda-\alpha p}$, $\tilde{p} = \frac{\lambda p}{\lambda-\alpha p}$, $\alpha p < \lambda$

$I_\alpha : H^{p,\lambda} \longrightarrow H^{\tilde{p},\lambda} \cap H^{p,\lambda+\alpha p'}$

One of the key results in the theory of Morrey Spaces is the action of the potential operator I_α.

Theorem 7.1. For $f \in L^{p,\lambda}$,

  (i) $||I_\alpha f||_{L^{\tilde{p},\lambda}} \leq c\,||f||_{L^{p,\lambda}}$, where $\tilde{p} = \lambda p/(\lambda-\alpha p)$, $\alpha p < \lambda \leq n$, $p > 1$.

  (ii) $||I_\alpha f||_{L^{p,\lambda-\alpha p}} \leq c\,||f||_{L^{p,\lambda}}$, again for $\alpha p < \lambda \leq n$, $p > 1$.

     And for $g \in H^{p,\lambda}$,

 (iii) $||I_\alpha g||_{H^{\tilde{p},\lambda}} \leq c\,||g||_{H^{p,\lambda}}$,

 (iv) $||I_\alpha g||_{H^{p,\lambda+\alpha p'}} \leq c\,||g||_{H^{p,\lambda}}$.

Theorem 7.1 (i) first appeared in [A3] in 1975. It can be considered the analogue of the Sobolev inequality in Morrey Spaces - and is often called the Morrey-Sobolev inequality. 6.1 (ii) is due to Peetre [P] in 1968. (iii) and (iv) are from [AX3].

Proof. For (i), we first estimate - via the Hedberg trick [He], for any $f \geq 0$

$$I_\alpha f(x) \leq c\,[M_{\lambda/p}f(x)]^{\alpha p/\lambda} \cdot [M_0 f(x)]^{1-\alpha p/\lambda}. \qquad (7.1)$$

In fact, upon writing $f = f_r + f^r$ as in Theorem 6.1, and $I_\alpha f = I_\alpha f_r + I_\alpha f^r = I_1 + I_2$,

$$I_1 = \sum_{k=0}^{\infty} \int_{2^{-k}r < |x-y| < 2^{-k+1}r} |x-y|^{\alpha-n} f(y)\, dy$$

$$\leq \sum_{k=0}^{\infty} (2^{-k}r)^{\alpha-n}(2^{-k+1}r)^n M_0 f(x) \leq C r^\alpha M_0 f(x),$$

© Springer International Publishing Switzerland 2015
D. Adams, *Morrey Spaces*, Applied and Numerical Harmonic Analysis,
DOI 10.1007/978-3-319-26681-7_7

and

$$I_2 = \sum_{k=1}^{\infty} \int_{2+k_r<|x-y|<2^{+k+1}r} |x-y|^{\alpha-n} f(y)\, dy$$

$$\leq \sum_{k=1}^{\infty} (2^{+k}r)^{\alpha-n} (2^{+k+1}r)^{n-\lambda/p} M_{\lambda/p} f(x)$$

$$= C\, r^{\alpha-\lambda/p} M_{\lambda/p} f(x),$$

for $0 < \alpha < \lambda/p$. Now choose

$$r = [M_{\lambda/p} f(x)/M_0 f(x)]^{p/\lambda},$$

resulting in (7.1).

Next, notice that $M_{\lambda/p} f \leq (M_\lambda f^p)^{1/p} \leq ||f||_{L^{p,\lambda}}$, hence raising (7.1) to the power $\tilde{p}$ and integrating over the ball $B(x_0, r)$ gives

$$\int_{B(x_0,r)} (I_\alpha f)^{\tilde{p}}\, dx \leq c\, ||f||_{L^{p,\lambda}}^{\frac{\alpha p}{\lambda} \cdot \tilde{p}} \cdot \int_{B(x_0,r)} (M_0 f)^p\, dx.$$

(i) now follows since M_0 is continuous on $L^{p,\lambda}$, Theorem 6.1.
For (ii), we again use (7.1) to get

$$\int_{B(x_0,r)} (I_\alpha f)^p\, dx \leq C\, ||f||_{L^{p,\lambda}}^{\frac{\alpha p}{\lambda} \cdot \tilde{p}} \cdot \int_{B(x_0,r)} (M_0 f)^{p(\lambda-\alpha p)/\lambda}\, dx.$$

Now apply Hölder's inequality and Theorem 6.1.
For (iii) and (iv), we use duality

$$\int I_\alpha g \cdot f\, dx = \int g \cdot I_\alpha f\, dx \leq ||g||_{H^{q,\lambda}} ||I_\alpha f||_{L^{\tilde{p},\lambda}}$$

$$\leq C\, ||g||_{H^{q,\lambda}} ||f||_{L^{\tilde{p},\lambda}}$$

where $q = \tilde{p}'$. Hence

$$||I_\alpha g||_{H^{\tilde{p}',\lambda}} \leq C\, ||g||_{H^{q,\lambda}} \tag{7.2}$$

which is just (iii) when one unravels the exponents.
Finally, (iv) follows the above pattern.   □

7.2 Wolff potentials associated with $||I_\alpha\mu||_{L^{p'}}$

Here our goal is to eventually estimate the size of the singular set for a Morrey potential

$$I_\alpha f, \quad f \in L_+^{p,\lambda},$$

i.e., the set where $I_\alpha f(x) = +\infty$. And to do this it is advantageous to get estimates on the Wolff potentials corresponding to this Morrey set up; i.e., we wish to estimate the Morrey norm of the potentials $I_\alpha\mu$ μ = Borel measure on $\mathbb{R}^n$ with compact support. To see what this might be, we first look at the Sobolev case. There one has the celebrated Wolff inequality

$$||I_\alpha\mu||_{L^{p'}}^{p'} \sim \int W_{\alpha,p}^\mu(x) \, d\mu(x) \tag{7.3}$$

where $W_{\alpha,p}^\mu$ is the Sobolev-Wolff potential:

$$W_{\alpha,p}^\mu(x) = \int_0^\infty [r^{\alpha p - n}\mu(B(x,r))]^{\frac{1}{p-1}} \frac{dr}{r},$$

$1 < p < n/\alpha$.

The proof (7.3) is contained in [AH], and what we do below will reduce to (7.3) when $\lambda = n$. But first, to see (7.3) one needs

$$||I_\alpha\mu||_{L^{p'}} \sim ||M_\alpha\mu||_{L^{p'}} \tag{7.4}$$

which is the case of Theorem 6.3 (i) for $\lambda = n$, $1 < p < \infty$. The next step is to invoke the elementary inequality

$$(M_\alpha\mu(x))^{p'} \le c \int_0^\infty [t^{\alpha-n} \, \mu(B(x,t))]^{p'} \frac{dt}{t}$$

and hence

$$\int (M_\alpha\mu)^{p'} \, dx \le c \int_0^\infty t^{(\alpha-n)p'} \int \mu(B(x,t))^{p'} \, dx \, \frac{dt}{t}. \tag{7.5}$$

But

$$\int \mu(B(x,t))^{p'} \, dx = \int \left(\int_{|x-y|<t} d\mu(y) \right)^{p'-1} \cdot \int_{|x-y'|<t} d\mu(y') \, dx$$

$$\le c \int \left(\int_{|y-y'|<2t} d\mu(y) \right)^{p'-1} t^n \, d\mu(y').$$

And with this into (7.5) gives the upper bound in (7.3).

For the lower bound, write

$$||I_\alpha \mu||_{L^{p'}}^{p'} = \int I_\alpha (I_\alpha \mu)^{\frac{1}{p-1}} \, d\mu \tag{7.6}$$

and estimate the non-linear potential $I_\alpha (I_\alpha \mu)^{\frac{1}{p-1}}$: it always exceeds

$$c \int_0^\infty r^{\alpha-n} \left(\int_{|x-y|<r} \int_{|x-z|<r} |y-z|^{\alpha-n} \, d\mu(z) \right)^{1/(p-1)} dy \, \frac{dr}{r}$$

which in turn exceeds

$$c \int_0^\infty r^{(\alpha-n)p'} \mu(B(x,r))^{1/(p-1)} \cdot r^n \, \frac{dr}{r}.$$

Hence

$$I_\alpha (I_\alpha \mu)^{\frac{1}{p-1}} (x) \geq c \cdot W_{\alpha,p}^\mu (x). \tag{7.7}$$

This with (7.6) gives the result (7.3).

Note, however, that the reverse of inequality (7.7) is in general false! (Take $\mu = $ Dirac measure of δ_0). This is basically why the result of T. Wolff was so startling in its time. In fact we see that the reverse of (7.7) holds for $p > 2 - \alpha/n$ only.

7.3   Wolff potentials associated with $||I_\alpha \mu||_{L^{p',\lambda}}$, $||I_\alpha \mu||_{H^{p',\lambda}}$

With this, we now look for the analogue of W in the Morrey theory, i.e., we wish to estimate

$$||I_\alpha \mu||_{L^{p',\lambda}}^{p'} \tag{7.8}$$

for $\lambda < n$.

Theorem 7.2. For $0 < \alpha < n$, $0 < \lambda < n$, $1 < p < n/\alpha$, $p' = p/(p-1)$ and μ a Borel measure on $\mathbb{R}^n$, it follows that

$$||I_\alpha \mu||_{L^{p',\lambda}}^{p'} \leq C \int W_{\alpha,p,\lambda}^\mu (y) \, d\mu(y); \quad \geq C \int W_{\alpha,p,\sigma}^\mu (y) \, d\mu(y), \quad \sigma > \lambda, \tag{7.9}$$

and

$$||I_\alpha \mu||_{H^{p',\lambda}}^{p'} \sim \inf_\omega \int W_{\alpha,p,\lambda}^{\mu,\omega} (y) \, d\mu(y) \tag{7.10}$$

where

$$W_{\alpha,p,\lambda}^{\mu}(x) \;=\; \int_0^\infty [r^{\alpha p-(n-\lambda)p-\lambda}\mu(B(x,r))]^{\frac{1}{p-1}}\,\frac{dr}{r},\tag{7.11}$$

and

$$W_{\alpha,p,\lambda}^{\mu,\omega}(x) \;=\; \int_0^\infty\left[\frac{r^{\alpha p}\mu(B(x,r))}{\int_{B(x,r)}\omega}\right]^{\frac{1}{p-1}}\,\frac{dr}{r},\tag{7.12}$$

and the infimum in (7.10) is over all $\omega\in A_1$ such that

$$\int \omega\,d\,\Lambda^{n-\lambda}\;\le\;1.\tag{7.13}$$

Proof. Notice that $W_{\alpha,p,n}^{\mu}$ agrees with $W_{\alpha,p}^{\mu}$ and that again (7.12) agrees with $W_{\alpha,p}^{\mu}$ when $\lambda=n$ provided we interpret $\int 1\,d\Lambda^0=1$.

Again, we begin by estimating $M_\alpha\mu$ as we did above. So here, we set

$$I(x,r,t) \;=\; \int_{B(x,r)}\mu(B(y,t))^{p'}\,dy$$

and then write

$$||M_\alpha\mu||_{L^{p'},\lambda}^{p'} \;\le\; \sup_{x,r>0} r^{\lambda-n}\left(\int_0^r+\int_r^\infty\right)t^{p'(\alpha-n)}I(x,r,t)\,\frac{dt}{t}\tag{7.14}$$

as the analogue of (7.5). But notice that

$$I(x,r,t) \;\le\; \int \mu(B(y,2t))^{p'-1}\cdot|B(x,r)\cap B(y,t)|\,d\mu(y).$$

So for the first integral in (7.14), we have that it does not exceed

$$c\int_0^\infty\left(t^{n+p'(\alpha-n)}t^{\lambda-n}\int\mu(B(y,2t))^{p'-1}\,d\mu(y)\right)\frac{dt}{t}.$$

For the second integral in (7.14), it does not exceed

$$\int_0^\infty\left(t^{p'(\alpha-n)+\lambda}\int\mu(B(y,2t))^{p'-1}\,d\mu(y)\right)\frac{dt}{t}$$

and the result follows as in the case $\lambda=n$.

Next, we look for the upper bound for (7.10). Here we write

$$||M_\alpha\mu||_{H^{p'},\lambda}^{p'} \;\le\; \inf_\omega\int\left(\int_0^\infty[t^{\alpha-n}\mu(B(x,t))]^{p'}\frac{dt}{t}\right)\omega(x)^{1-p}\,dx.$$

So now, set

$$J(t) = \int \mu(B(x,t))^{p'} \, \omega(x)^{1-p'} \, dx$$

and

$$J(t) \leq \int \left(\int_{B(y,t)} \mu(B(x,t))^{p'-1} \omega(x)^{1-p'} \, dx \right) d\mu(y)$$

$$\leq c \, \mu(B(y,2t))^{p'-1} \, t^n \left(t^{-n} \int_{B(y,t)} \omega(x) \, dx \right)^{1-p'}$$

because $\omega \in A_1$ and hence in A_p. So the upper bound holds.

The lower bounds are a bit easier (and do not depend on Theorem 6.3 (i) or (ii)). One again proceeds by getting a lower bound on the corresponding non-linear potential as we did earlier. But since those lower bounds do not play a role in these lecture notes, we refer the reader to[AX3] and [AX7] for details.   □

7.4   A "Morrey bridge" to C^α

We end this Chapter with a curious observation, namely

$$H^{p,\lambda} \xrightarrow{I_{n-\lambda}} VL^{p,\lambda} \xrightarrow{I_\alpha} C^{\alpha-\lambda/p}$$

which makes the Morrey Space a bridge to the Holder continuous functions from $H^{p,\lambda}$ to $C^{\alpha-\lambda/p}$.

Theorem 7.3. For $g \in H^{p,\lambda}$

$$||I_{n-\lambda}g||_{L^{p,\lambda}} \leq c \, ||g||_{H^{p,\lambda}}$$

for some constant independent of g.

Proof. For $g \geq 0$,

$$\int_{B(x,r)} (I_{n-\lambda}g(y))^p \, dy = \int_{B(x,r)} \left(\int |y-z|^{-\lambda} g(z) \, dz \right)^p \, dy$$

$$\leq c \int_{B(x,r)} \left(\int |y-z|^{-\lambda p'} \omega(z) \, dz \right)^{p-1} \left(\int g(z)^p \, \omega(z)^{1-p} \, dz \right) dy$$

$$\leq c \int g(z)^p \, \omega(z)^{1-p} \, dz \cdot \int_{B(x,r)} (I_{n-\lambda p'} \, \omega(y))^{p-1} \, dy.$$

And now we need a lemma, whose proof we put in the Notes at the end of this chapter.

Lemma 7.4. If $\omega \in L^1_+(\Lambda^{n-\lambda})$, then

$$\int \omega^{\frac{n}{n-\lambda}} \, dy \leq c\left(\int \omega \, d\Lambda^{n-\lambda}\right)^{\frac{n}{n-\lambda}}$$

for some constant independent of ω.

Thus continuing the argument above, we have that if $\omega \in L^1_+(\Lambda^{n-\lambda})$, then $I_{n-\lambda p'}\omega \in L^{(p-1)n/\lambda}$ by the standard Sobolev inequality. Hence since we require $\int \omega \, d\Lambda^{n-\lambda} \leq 1$, we can write

$$\int_{B(x,r)} (I_{n-\lambda-p'}\omega)^{p-1} \, dy \leq C \, r^{n-\lambda}\left(\int_{B(x,r)} (I_{n-\lambda p'} \, \omega)^{n(p-1)/\lambda}\right)^{\lambda/n}$$

$$\leq C \, r^{n-\lambda}\left(\int \omega^{\frac{n}{n-\lambda}} \, dy\right)^{\frac{n-\lambda}{n}} \leq C \, r^{n-\lambda}.$$

And finally, the fact that $I_{n-\lambda}$ actually maps into $VL^{p,\lambda}$ is a consequence of the fact that C_0 is dense in $H^{p,\lambda}$. And the map $I_\alpha : L^{p,\lambda} \longrightarrow C^{\alpha-\lambda/p}$ is, of course the standard Morrey's Lemma. $\qquad\square$

7.5 Notes

7.5.1 Proof of Lemma 7.4

The proof of Lemma 7.4 is taken from [OV]. For $f \geq 0$.

$$\int f(x) \, dx = \int_0^\infty \mathscr{L}^n \, (f > t) \, dt$$

$$\leq \int_0^\infty \Lambda^d(f > t)^{n/d} \, dt$$

$$= \frac{n}{d} \int_0^\infty \Lambda^d(f > r^{n/d}) \, r^{n/d} \, \frac{dr}{r}$$

$$\leq c\left(\int f^{d/n} \, d\Lambda^d\right)^{n/d-1} \cdot \int_0^\infty \Lambda^d(f > r^{n/d}) \, dr$$

$$\leq c\left(\int f^{d/n} \, d\Lambda^d\right)^{n/d}.$$

because

$$r \, \Lambda^d (f > r^{n/d}) \leq \int_{[f^{d/n} > r]} f^{d/n} \, d\Lambda^d,$$

by Holder's inequality - first for interior dyadic Hausdorff capacity, and then for Λ^d.

7.5.2 $I_\alpha : L^{1,\lambda} \longrightarrow L_w^{\tilde{1},\lambda}, \ \tilde{1} = \lambda/(\lambda - \alpha), \ 0 < \alpha < \lambda$

From [A3], one also has that

$$I_\alpha : L^{1,\lambda} \longrightarrow L_w^{\tilde{1},\lambda}$$

where $\tilde{1} = \lambda/(\lambda - \alpha), \ 0 < \alpha < \lambda$ - the corresponding weak type result for $p = 1$ of Theorem 6.1.

Chapter 8
Singular Integrals on Morrey Spaces

8.1 $T : L^{p,\lambda} \longrightarrow L^{p,\lambda}$ and $T : H^{p,\lambda} \longrightarrow H^{p,\lambda}$; $1 < p < \infty$, $0 < \lambda < n$

Here we want to consider the action of the Calderon-Zygmund singular integrals on the Morrey Spaces. This has a brief history, namely due to Stampacchia and later Peetre. What was known up to 1969 is summarized in the paper [Pe1]. But their approach was really one via interpolation: knowing that the singular operator T maps $L^p \to L^p$ for different p (and $\lambda = n$), allows that T then maps an L^p for an intermediate p into $L^{p,\lambda}$, $\lambda = \frac{p}{p_0} n(1-\theta) + \frac{p}{p_1} n\theta$. But one thing he could not do was have $L^{p,\lambda}$ in the domain space. Later in Chapter 11, we will remove this restriction, but there are some relevant counterexamples that need to be addressed.

But our theory of Chapter 5 can now be used to show that such singular operators map $L^{p,\lambda} \to L^{p,\lambda}$ directly. This is an easy consequence of A_p weight theory in HA.

Theorem 8.1. Let T be a Calderon-Zygmund singular integral operator, or any linear operator that maps weighted L^p into weighted L^p, $\omega \in A_p$. Then

(i) $T : L^{p,\lambda} \longrightarrow L^{p,\lambda}$
   and
(ii) $T : H^{p,\lambda} \longrightarrow H^{p,\lambda}$
   $1 < p < \infty$, $0 < \lambda < n$

Proof. Take $\omega \in A_1$ with $\int \omega \, d\Lambda^{n-\lambda} \leq 1$. Then from [St2] or [To], we have $\omega^{1-p} \in A_p$. Hence

$$\int |Tf|^p \, \omega^{1-p} \, dx \leq c \int |f|^p \, \omega^{1-p} \, dx,$$

© Springer International Publishing Switzerland 2015
D. Adams, *Morrey Spaces*, Applied and Numerical Harmonic Analysis,
DOI 10.1007/978-3-319-26681-7_8

by the L^p weight theory. Consequently,

$$||Tf||_{H^{p,\lambda}} \leq C||f||_{H^{p,\lambda}}.$$

This gives (ii). For (i), we just invoke duality

$$\int Tf \cdot g \, dx = \int fT^*g \, dx \leq ||f||_{L^{p,\lambda}} ||T^*g||_{H^{p',\lambda}}$$

$$\leq c||f||_{L^{p,\lambda}} ||g||_{H^{p',\lambda}};$$

$T^* = $ adj of T.   □

This works well for operators $T : T^* = T$ (self adjoint); e.g.

$$Tf(x) = p.v. \int K(x-y)f(y) \, dy.$$

Other operators can also be addressed - e.g., so-called maximal singular integral operators. Again, the reader should consult [St2] or [To].

Chapter 9
Morrey-Sobolev capacities

9.1 Definitions and simple properties for $C_{\alpha,p}(\cdot)$

<u>Sobolev capacities.</u> Here we refer the reader to [AH] for the theory of the capacities $C_{\alpha,p}$ and $\dot{C}_{\alpha,p}$ and in fact for the general theory of capacities based on potentials of functions in Lebesgue spaces; see also Meyers [M2] 1970.

For our purposes, let K be a compact set in $\mathbb{R}^n$, and set

$$C_{\alpha,p}(K) = \inf\{\|f\|_{L^p}^p : G_\alpha f \geq 1 \text{ on } K; f \geq 0\}$$
$$\dot{C}_{\alpha,p}(K) = \inf\{\|f\|_{L^p}^p : I_\alpha f \geq 1 \text{ on } K; f \geq 0\}.$$

These two set functions are sometimes referred to as the L^p- Bessel and L^p- Riesz capacities, respectively. $\dot{C}_{\alpha,p}$ is the homogeneous version of $G_\alpha f$. Here G_α is the Bessel kernel of order α; i.e., the Fourier transform of G_α satisfies

$$\widehat{G_\alpha}(\zeta) = (1 + |\zeta|^2)^{-\alpha/2}, \ \alpha > 0.$$

This is chosen because it is easy to see that

$$G_\alpha f(x) = \int_{\mathbb{R}^n} G_\alpha(x-y) f(y) \, dy \left(I_\alpha f(x) = \int_{\mathbb{R}^n} |x-y|^{\alpha-n} f(y) \, dy \right) \quad (9.1)$$

is an integral representation of a Sobolev function when $\alpha = m \in \mathbb{N}$. In fact,

$$\|f\|_{L^p}^p \sim \|u\|_{\text{Sobolev}}^p = \|u\|_{L^p}^p + \|D^m u\|_{L^p}^p,$$

© Springer International Publishing Switzerland 2015
D. Adams, *Morrey Spaces*, Applied and Numerical Harmonic Analysis,
DOI 10.1007/978-3-319-26681-7_9

$u = G_m f$. D^β represents derivatives of order $\beta \in \mathbb{N}$. Further, it is well known that there is a constant c such that

$$\frac{1}{c}\, \dot{C}_{\alpha,p}(K) \leq C_{\alpha,p}(K) \leq a\left(\dot{C}_{\alpha,p}(K) + \dot{C}_{\alpha,p}(K)^{\frac{n}{n-\alpha p}} \right)$$

for $\alpha p < n$. Also $\dot{C}_{\alpha,p}(\cdot) \equiv +\infty$ when $\alpha p = n$.

Furthermore, the min-max theorem implies that both $C_{\alpha,p}$ and $\dot{C}_{\alpha,p}$ have a dual version, e.g.

$$\dot{C}_{\alpha,p}(K)^{1/p} = \sup\{||\mu||_1 : \operatorname{supp}\mu \subset K \ \& \ ||I_\alpha \mu||_{L^{p'}} \leq 1\}$$

$||\mu||_1 = \mu(\mathbb{R}^n)$. And because of the celebrated Wolff inequality (7.4), we can define an equivalent capacity in terms of $W^\mu_{\alpha,p}$. In fact,

$$\dot{C}_{\alpha,p}(K) \sim \inf\{||\mu||_1 : W^\mu_{\alpha,p} \leq 1, \ \operatorname{supp}\mu \subset K\}.$$

See [AH].

Other properties of the capacity $\dot{C}_{\alpha,p}$ are:

$$\dot{C}_{\alpha,p}(B(x,r)) \sim r^{n-\alpha p}, \ \alpha p < n$$

and

$$C_{\alpha,p}(B(x,r)) \sim \begin{cases} r^{\alpha p - n}, & \alpha p < n \\ (\log 1/r)^{1-p}, & \alpha p = n. \end{cases}$$

Also, $C_{\alpha,p}(K) > 0$ for all compact sets (non empty) when $\alpha p > n$.

9.2   Definitions and simple properties for $C_\alpha(\cdot\,; X), X = L^{p,\lambda}$ or $H^{p,\lambda}$

The main purpose of this chapter is to introduce the capacities associated with potentials of functions in the Morrey Spaces and ideally to produce analogues of all the results for these capacities as for the Sobolev capacities. In [AX2], the authors introduced and studied such capacities - those associated with potentials of functions in $H^{p,\lambda}$, Type I capacities, and for potentials of functions in $L^{p,\lambda}$, Type II capacities.

So we begin by writing

$$C_\alpha(K;X) = \inf\{||f||_X^p : I_\alpha f \geq 1 \text{ on } K, f \geq 0\}$$

where X will be taken to be any of $H^{p,\lambda}, L^{p,\lambda}$, or $VL^{p,\lambda}$. Here K is a compact subset of $\mathbb{R}^n$. (Note that by standard methods, these capacities can be extended to all Borel set E, and even all analytic sets; see [AH]).

And looking for a dual capacity, we set

$$Cap_\alpha(K;X) \equiv \sup\{\|\mu\|_1 : \operatorname{supp}\ \mu \subset K\ \&\ \|I_\alpha\mu\|_{X^*} \le 1\}$$

Here X^* is the dual space to X. So to have a capacity and it dual in the theory, we generally want to confine X to either $VL^{p,\lambda}$ or $H^{p,\lambda}$. But in any case, it is not hard to see that

$$C_\alpha(K; L^{p,\lambda}) \sim C_\alpha(K; VL^{p,\lambda})$$

because if $f \in L^{p,\lambda}$, $\varphi_\epsilon * f$ is continuous for φ_ϵ an approximation to the identity and

$$I_\alpha(\varphi_\epsilon * f) = I_\alpha\ \varphi_\epsilon * f \longrightarrow I_\alpha f$$

due to the estimate

$$\varphi_\epsilon * f \le c \cdot M_0 f, \quad \text{for } \epsilon > 0.$$

And again the min-max theorem gives

$$C_\alpha(K;X) = Cap_\alpha(K;X^*).$$

See [AH] or [AX2] for details.

Our main interest lies with

Theorem 9.1. There is a constant $a > 0$ such that

$$\frac{1}{a} \cdot \frac{C_\alpha(K \cap B(x,r); H^{p,\lambda})}{r^{(n-\lambda)p+\lambda-\alpha p}} \le \frac{C_\alpha(K \cap B(x,r); L^p)}{r^{n-\alpha p}}$$

$$\le a \cdot \frac{C_\alpha(K \cap B(x,r); L^{p,\lambda})}{r^{\lambda - \alpha p}}. \tag{9.2}$$

Here clearly, we are also writing $C_\alpha(K; L^p)$ for $C_{\alpha,p}(K)$.

Proof. Notice that (9.2) determines the relative strength of these three capacities in the sense: C is stronger than C', if $C(K) = 0$ always implies $C'(K) = 0$. Furthermore, it is easy to see that all three are capacities in the sense of Meyers (see Chapter 3). So it is easy enough to prove (9.2) for $K \subset B(0,1)$.

To get (9.2), suppose $Cap_\alpha(K \cap B_r; VL^{p',\lambda})$ is positive, then there is a measure μ supported by $K \cap B_r$ such that

$$\int_{B(x,r)} (I_\alpha\mu)^{p'}\ dy \le r^{n-\lambda}, \quad r > 0.$$

Then the setting $\gamma = r^{\lambda-n} \cdot \mu$ and $\omega = r^{\lambda-n} \chi_{B(x,r)}$ gives

$$\int (I_\alpha \gamma)^{p'} \omega^{1-p'} \, dy \leq 1,$$

and then

$$r^{\lambda-n} Cap_\alpha(K \cap B_r; VL^{p',\lambda}) \leq c \cdot Cap_\alpha(K \cap B_r; H^{p,\lambda}).$$

Or via the min-max theorem

$$r^{(\lambda-n)p} \, C_\alpha(K \cap B_r; H^{p,\lambda}) \leq c \cdot C_\alpha(K \cap B_r; L^{p,\lambda}).$$

A similar calculation gives

$$C_\alpha(K \cap B_r; H^{p,\lambda}) \leq c \cdot r^{(n-\lambda)(p-1)} \cdot C_\alpha(K \cap B_r; L^p).$$

$\square$

Remark These inequalities suggest the need to define a reasonable notion of "thinness" associated with each of these capacities - that for $C_\alpha(\cdot, L^p) = \dot{C}_{\alpha,p}$ is of course well known and extensively studied in [AH, HKM], and more recently in [BB]; as it is also for $C_{\alpha,p}(\cdot)$. We will, however, postpone such a discussion of such thinness (of a set) and the relations among them for another time - in fact another open question for the reader! However, it is certain that the Wolff potentials in these cases will play a fundamental role (cf. Chapter 7).

9.3 $C_\alpha(B(x,r); L^{p,\lambda})$

Next, we calculate the capacity of a ball.

Theorem 9.2.

(a) $C_\alpha(B(x,r); L^{p,\lambda}) \sim r^{\lambda-\alpha p}, \quad \alpha p < \lambda,$

(b) $C_\alpha(B(x,r); L^{p,\lambda}) \sim (\log 1/r)^{-p}, \quad \alpha p = \lambda.$ (9.3)

And consequently by countable subadditivity

$$C_\alpha(K; L^{p,\lambda}) \leq c \cdot \Lambda^{\lambda-\alpha p}(K), \quad \alpha p < \lambda. \tag{9.4}$$

Proof. For (a) if $X_{B(x,r)}$ is the usual characteristic function of the ball $B(x,r) \subset \mathbb{R}^n$, then

$$\|X_{B(x_0,r_0)}\|_{L^{p,\lambda}}^{p'} = \sup_{\substack{r>0 \\ x_0 \in \mathbb{R}^n}} |B(x_0,r_0) \cap B(x,r)|$$

$$\sim \sup_{r>0} \begin{cases} r^\lambda, & r < 2r_0 \\ r^{\lambda-n}r_0^n, & r > 2r_0 \end{cases} = c \cdot r_0^\lambda.$$

Also

$$I_\alpha * X_{B(x,r)}(y) \geq c \cdot r^\alpha$$

when $y \in B(x,r)$. Thus

$$C_\alpha(B(x,r); L^{p,\lambda}) \leq \|X_{B(x,r)} \cdot r^{-\alpha}\|_{L^{p,\lambda}}^p$$
$$\sim r^{\lambda-\alpha p}, \quad \alpha p < \lambda.$$

For the lower bound, we use Theorem 7.1(i):

If

$$I_\alpha f \geq X_{B(x,r)}$$

then

$$\|X_{B(x,r)}\|_{L^{p,\lambda}}^{\bar{p}} \leq C_\alpha(B(x,r); L^{p,\lambda})^{\bar{p}/p}.$$

But again since the left side exceeds $c \cdot r^\lambda$ the result follows.
   For (b), take $(r < 1)$

$$f(x) = \begin{cases} 0, & x \in \mathbb{R}^n \backslash B(x_0,1), \\ \left(\dfrac{r}{|x-x_0|}\right)^\alpha, & x \in B(x_0,1) \backslash B(x_0,r), \\ 1, & x \in B(x_0,r). \end{cases}$$

then,

$$I_\alpha f(x) = \int_{B(x_0,r)} \cdots + \int_{B(x_0,1)\backslash B(x_0,r)} \cdots + \int_{\mathbb{R}^n \backslash B(x_0,1)} \cdots$$

$$\geq r^\alpha \int_{B(x_0,1)\backslash B(x_0,r)} |y-x|^{\alpha-n}|y-x_0|^{-\alpha}\, dy$$

$$\geq c\, r^\alpha \int_{B(x_0,1)\backslash B(x_0,r)} |y-x_0|^{-n}\, dy \geq c\, r^\alpha \log\frac{1}{r}.$$

And

$$\|f\|_{L^{p,\lambda}}^p \leq \sup_{\substack{t > 0 \\ x_0 \in \mathbb{R}^n}} \left(|B(x,t) \cap B(x_0,r)| + \int_{B(x,t) \cap [B(x_0,1) \setminus B(x_0,r)]} \left(\frac{r}{|x - x_0|} \right)^{\alpha p} dy \right)$$

$$\leq c \cdot r^\lambda.$$

Hence

$$\left\| \frac{f}{r^\alpha \log 1/r} \right\|_{L^{p,\lambda}}^p \leq c \, (\log 1/r)^{-p}.$$

For a lower bound, we use the following extension of Theorem 7.1 (i):

Lemma 9.1. For $\alpha p = \lambda, 1 < p = \lambda/\alpha$,

$$\|I_\alpha f\|_{\text{BMO}} \leq c \|f\|_{L^{p,\lambda}}.$$

Here BMO is the John-Nirenberg space of functions of bounded mean oscillation $\mathscr{L}^{p,0}$; see Introduction and Lemma 2.1 of Chapter 2. There the BMO quasi-norm was denoted by $[f]_* = [f]_{*,\mathbb{R}^n} = \|f\|_{\text{BMO}}$.

So, continuing with our proof of (b), we write via Lemma 9.1, (supp $f \subset B_0$)

$$\sup_{B \subset \mathbb{R}^n} |B|^{-1} \int_B \exp \left(\beta \frac{|I_\alpha f - (I_\alpha f)_B|}{\|f\|_{L^{p,\lambda}}} \right) dx \leq c$$

for some $\beta > 0$. But working on the fixed ball B_0, we have

$$\int_{B_0} I_\alpha f \, dx \leq c \, \|f\|_{L^{p,\lambda}}.$$

Hence

$$\int_{B_0} \exp(\beta \, \|f\|_{L^{p,\lambda}}^{-1} \, I_\alpha f) \, dx \leq c.$$

Thus for $I_\alpha f \geq 1$ on $B(x,r) \subset B_0$, it follows that

$$\exp(\beta \, \|f\|_{L^{p,\lambda}}^{-1}) \cdot |B(x,r)| \leq c$$

which in turn gives

$$\|f\|_{L^{p,\lambda}}^p \geq c \left(\log \frac{1}{r} \right)^{-p}.$$

$\square$

9.4 $C_\alpha(B(x,r); H^{p,\lambda})$ and failure of CSI for $C_\alpha(\cdot;\ L^{p,\lambda})$

Now we could continue with a discussion of the properties of the Type I capacities here, but that is not needed for our general goal of these notes: applications to PDE-Chapters 15 and 16. Suffice it to say that the best, so far, analogue of Theorem 9.2 for Type I capacities is:

Theorem 9.3. For $1 < p < \infty$, $0 < \lambda < n$,

$$(i) \qquad C_\alpha(B(x,r); H^{p',\lambda}) \le c \cdot r^{(n-\alpha)p-\lambda},$$

and

$$(ii) \qquad C_\alpha(B(x,r); H^{p',\lambda}) \ge c \cdot r^{(n-\alpha)p-\lambda},$$

for $p > \max\left(\dfrac{\lambda}{n-\alpha}, \dfrac{n-\lambda}{\alpha}\right)$.

The reader is referred to [AX2] for a further discussion of these Type I capacities. But one feature in [AX2] for both Type I and Type II capacities is whether or not these capacities satisfy a "Maz'ya strong type" inequality. For $\dot{C}_{\alpha,p} = \dot{C}_\alpha(\cdot; L^p)$, it known that there is a constant $c > 0$ independent of f such that

$$\int_0^\infty \dot{C}_{\alpha,p}\left([I_\alpha f > t]\right)\, dt^p \equiv \int (I_\alpha f)^p\, d\dot{C}_{\alpha,p} \le c\|f\|_{L^p}^p, \tag{9.5}$$

for $f \ge 0$, (A similar inequality also holds for $C_{\alpha,p}$). $\alpha p < n$. Equation (9.5) is the strong type analogue of the trivial estimate

$$\dot{C}_{\alpha,p}([I_\alpha f > t]) \le \frac{1}{t^p}\|f\|_{L^p}^p,$$

a weak type inequality. Such is the case for Type I capacities, but Type II, we have

Theorem 9.5. (i) For $0 < \alpha < n$, $0 < \lambda < n$, $\gamma > 0$, $1 < p < \lambda/\alpha$, there is a function $f \in L^{p,\lambda}$ such that

$$\int_0^\infty [C_\alpha([I_\alpha f > t]; L^{p,\lambda})\, t^p]^{\gamma/p} \frac{dt}{t} = +\infty, \tag{9.6}$$

and (ii) there is a constant c independent of f such that for $0 < \alpha < \lambda < n$, $1 < p < \lambda/(\lambda - \alpha)$ and $f \in H^{p',\lambda}$,

$$\int (I_\alpha f)\, dC_\alpha(\cdot\,; H^{p',\lambda}) \le c\,\|f\|_{H^{p',\lambda}}. \tag{9.7}$$

We refer the reader to [AX2]. Note, however, the weak type inequality is again trivial

$$C_\alpha([I_\alpha f > t]; L^{p,\lambda}) \leq \frac{1}{t^p}\|f\|^p_{L^{p,\lambda}},$$

which corresponds to $\gamma = +\infty$ in (9.6). Thus one gets the idea that the capacities $C_\alpha(\cdot\,;\,L^{p,\lambda})$ are just to "big" to support a strong type inequality.

9.5   Notes

9.5.1   *Weighted capacity vs. Choquet Integrals*

As we mentioned in Chapter 3, some effort has been made to establish Radon-Nikodym type results for weighted capacities. So here, we draw the readers' attention to such for weighted $\dot{C}_{\alpha,p}$, for example. The general question is whether or not we can write something like

$$\dot{C}^w_{\alpha,p}(K) \sim \int_K w\, d\dot{C}_{\alpha,p}$$

when

$$C^w_{\alpha,p}(K) = \inf\{\|f\|^p_{L^p(w)} : I_\alpha f \geq 1 \text{ on } K, f \geq 0\}$$

with

$$\|f\|^p_{L^p(w)} = \int |f(y)|^p\, w(y)\, dy?$$

Partial results along these lines can be found in [A6]. And such questions are wide open for the Morrey-Sobolev capacities - especially in that capacities of Type I are already defined via weighted norm.

9.5.2   *Relations between $C_{\alpha,p}$ and $C_{\beta,q}$ via Morrey Theory*

An early question concerning the capacities $C_{\alpha,p}$ and $\dot{C}_{\alpha,p}$ was the relations between them - see [AH] for such. But an interesting point we can make here is that one can derive these relations, i.e., esp $C_{\alpha,p} << C_{\beta,q}$ for $\alpha p = \beta q$, $q < p$, using the Morrey-Sobolev theory, esp. inequality (i) of Theorem 7.1; see also [A3]. Using this approach is somewhat better than the original approach via boundedness principle for the non-linear potentials $I_\alpha(I_\alpha\mu)^{p'-1}$; see [AH]. Such a principle may not be available for capacities defined using more general kernels, say, for example, the Heat Kernel

$$\Gamma_\alpha(x,t) = \begin{cases} c_\alpha t^{\frac{\alpha-n}{2}} e^{-|x|^2/t}, & t > 0 \\ 0 & t \le 0, \end{cases}$$

and the corresponding Heat L^p- capacities , now even Heat $L^{p,\lambda}$ - capacities. This point is made because one can easily connect the L^p theory capacities with $L^{p,\lambda}$ theory capacities via : if μ is an extremal measure for the dual capacity $\dot{C}_{\alpha,p}$, then one can show that $I_\alpha\mu$ belongs to the Morrey space $L^{p',\alpha p}, \alpha p < n$. An extremal measure here is one that gives equality in the definition. Further, it is often the externals measures or functions that come up when dealing with elliptic PDE.

9.5.3   Speculations and parabolic capacities

Finally, since we have referred to parabolic results just above, it would seem that a natural question that arises here is indeed, the parabolic analogue of Morrey-Sobolev capacities or just Morrey theory in general - how much of what we present in these notes is true in the parabolic case? And this extends to treating parabolic equation analogues of Chapters 15 and 16.

Chapter 10
Traces of Morrey Potentials

10.1 $||I_\alpha f||_{L^q(\mu)} \leq c\,||f||_{L^p}$

We now come to the question of the size of the sets where $I_\alpha f$, $f \in L_+^{p,\lambda}$, cannot be defined, i.e., where $I_\alpha f(x) = +\infty$, $f \geq 0$ and $f \in L^{p,\lambda}$. The trace of such a potential is the set where the $I_\alpha f(x) < \infty$. Such sets of course can be measured with the capacities $\dot{C}_\alpha(\cdot\ ;\ L^{p,\lambda})$, $\alpha p \leq \lambda$, but putting this in terms of Hausdorff capacity is generally the acceptable way to get a more classical understanding of these potentials. But we start by reviewing the Sobolev case: $I_\alpha f$, $f \geq 0$ and $f \in L^p$. Here it has been known since the early 70s that such potentials are finite Λ^d - a.e. for $d > n - \alpha p$, $\alpha p < n$. In fact, we have, from [A1].

Theorem 10.1. There is a constant c_0 independent of f such that

$$||I_\alpha f||_{L^q(\mu)} \leq c_0||f||_{L^p} \tag{10.1}$$

for $q = \dfrac{dp}{n - \alpha p}$, $d > n - \alpha p$, $1 < p < n/\alpha$.
    iff

$$\mu(B(x,r)) \leq c_1 r^d \tag{10.2}$$

for all $x \in \mathbb{R}^n$ and $r > 0$. Here μ is a Borel measure on $\mathbb{R}^n$ and the integral in (10.1) is taken with respect to μ.

Proof. We give a proof here quite different from the original [A1]. We appeal to the Wolff potentials $\dot{W}_{\alpha,p}^\mu$ of Chapter 7. In particular, that theory implies that if $\dot{W}_{\alpha,p}^\mu$ is bounded then $\dot{C}_{\alpha,p}(\text{supp}\ \mu) > 0$, in fact

$$C_{\alpha,p}(\text{supp}\ \mu) \geq c_2||\mu||_1 \tag{10.3}$$

© Springer International Publishing Switzerland 2015
D. Adams, *Morrey Spaces*, Applied and Numerical Harmonic Analysis,
DOI 10.1007/978-3-319-26681-7_10

with $c_2 = c(\sup \dot{W}^{\mu}_{\alpha,p})^{p-1}$, for some constant c depending only on n, α, p. This follows from the min-max theorem - see [AH] and the Wolff inequality (7.3). Thus it is clear that if μ satisfies (10.2) and compact support with $d > n - \alpha p$, then $\dot{W}^{\mu}_{\alpha,p}$ is bounded and then Frostman's Theorem (Theorem 3.1) gives

$$C_{\alpha,p}(\text{supp } \mu) \geq c \cdot \Lambda^d(\text{ supp } \mu)^{(n-\alpha p)/d}$$

Thus we expect $I_\alpha f$, $f \geq 0$, $f \in L^p$, to be finite on such a set; $\alpha p < n$. This motivates our estimate (10.1) - which we now prove in detail.

So let $I_\alpha f \geq 1$ on $K \subset\subset \mathbb{R}^n$, $f \geq 0$ and smooth, then

$$\mu(K) \leq \int_K I_\alpha f \, d\mu = \int f \cdot I_\alpha \mu^K \, dx$$

$$\leq ||f||_{L^p} \, ||I_\alpha \mu^K||_{L^{p'}} \leq c \cdot \dot{C}_{\alpha,p}(K)^{1/p} \left(\int W^{\mu^K}_{\alpha,p} \, d\mu^K \right)^{1/p'}. \qquad (10.4)$$

Here μ^K is μ restricted to K. And now we estimate the Wolff potential as

$$\int_0^\infty [r^{\alpha p - n} \mu(B(x,r))]^{\frac{1}{p-1}} \frac{dr}{r} = \int_0^\delta \cdots + \int_\delta^\infty \cdots$$

$$\int_0^\delta \cdots \leq c \int_0^\delta [r^{\alpha p - n + d}]^{\frac{1}{p-1}} \frac{dr}{r} = c \, \delta^{\frac{\alpha p - n + d}{p-1}}, \, d < n - \alpha p$$

And

$$\int_\delta^\infty \cdots \leq c \, ||\mu||_1^{\frac{1}{p-1}} \delta^{\frac{\alpha p - n}{p-1}}, \, \alpha p < n.$$

Hence choosing $\delta = ||\mu||_1^{1/d}$ gives the estimate

$$\dot{W}^{\mu}_{\alpha,p}(x) \leq c \, ||\mu||_1^{\frac{\alpha p - n + d}{d(p-1)}}.$$

Putting this back into (10.4), now with $\mu = \mu^K$, we have

$$\mu(K)^{\frac{n-\alpha p}{dp}} \leq c \, C_{\alpha,p}(K)^{1/p}.$$

Finally let $K = E_t = [I_\alpha f \geq t]$, then it follows that

$$\mu(E_t) \leq \left(\frac{c}{t} \, ||f||_{L^p} \right)^q \qquad (10.5)$$

$d > n - \alpha p$, $1 < p < n/\alpha$, with $q = dp/(n - \alpha p)$. But (10.5) is just the weak type (p, q) inequality - so it is clear that the Marcinkewicz Interpolation Theorem gives (10.1); see [St2] or [To].

For the converse, (10.1) clearly implies

$$\mu(K)^{1/q} \leq c \, \dot{C}_{\alpha,p}(K)^{1/p}.$$

So if $K = B(x,r)$, $\dot{C}_{\alpha,p}(B(x,r)) \leq c \, r^{n-\alpha p}$, $\alpha p < n$, we easily get (10.2).   □

10.2 $\|I_\alpha f\|_{L^q(\mu)} \leq c_0 \|f\|_{L^{p,\lambda}}$

The question now is: what kind of trace estimates can we get for Riesz-Morrey potentials when $\lambda < n$? The answer is rather surprising for again it turns out that we must still assume $d > n - \alpha p$ even though $\alpha p \leq \lambda$. Recall that $I_\alpha f(x)$ is bounded - even Hölder continuous - when $\alpha p > \lambda$, by Morrey's Lemma (1.3), $f \in L^{p,\lambda}$, $\lambda < n$. Thus there is a gap between (9.4) and the known lower bound: $C_{\alpha,p}(K) \geq c \cdot \Lambda^d(K)^{(n-\alpha p)/d}$, $d > n - \alpha p$; cf. (9.2). Notice also, that $f_0(y) = |y|^{-\lambda/p} \in L^{p,\lambda}(B(0,1))$ and then $I_\alpha f_0(x) \sim |x|^{\alpha - \lambda/p}$, hence $I_\alpha f_0 \notin L^q_{loc}(\mu)$ when $q \geq dp/(\lambda - \alpha p)$, $\alpha p < \lambda$, and μ a d-measure, i.e., (10.2) holds.

A more complete treatment of trace estimates of Riesz-Morrey potentials is contained in [AX7]. We begin here with:

Theorem 10.2. Let μ be a compactly supported Borel measure on $\mathbb{R}^n$ satisfying (10.2), then if $1 < p < \lambda/\alpha$, $d > n - \alpha p$, $\alpha > 0$, there is a constant $c_0 > 0$ independent of f such that

$$\left(\int_B |I_\alpha f|^{q_0} \, d\mu \right)^{1/q_0} \leq c_0 \|f\|_{L^{p,\lambda}(B)} \tag{10.6}$$

for all $q_0 \in [p, (dp - (n - \lambda)p)/(\lambda - \alpha p))$. Here B is a fixed ball in $\mathbb{R}^n$.

The proof relies on

Lemma 10.3. For $f \in L^{p,\lambda}_+(B)$, μ as above, then

(a) $$\left(\int_B (I_\alpha f)^s d\mu \right)^{1/s} \leq c_1 \|f\|_{L^{p,\lambda}(B)}$$
          $1 \leq s < p$ and $d > n - \alpha p$, $\alpha p < \lambda$.

(b) $$I_\alpha f(x) \leq c_2 \|f\|_{L^{p,\lambda}(B)}, \qquad \forall x \in B, \quad \alpha p > \lambda.$$

Proof of (a): We use duality, setting $E_t = [I_\alpha f > t]$ and $F = B \cap E_t$ and $\mu^F = \mu \llcorner F$. Then

$$t\mu(F) \leq \int_B I_\alpha f \, d\mu^F = \int_B f \cdot I_\alpha \mu^F \, dx$$

$$\leq \|f\|_{L^{p,\lambda}(B)} \|I_\alpha \mu^F\|_{H^{p',\lambda}(B)}.$$

Here

$$\|I_\alpha \mu^F\|_{H^{p',\lambda}(B)} = \inf_w \int_B (I_\alpha \mu^F)^{p'} w^{-\frac{1}{p-1}} \, dx$$

where $w \in A_1$ and $\int_B w \, d\Lambda^{n-\lambda} \le 1$. Upon choosing $w \equiv 1$, we have

$$\int_{|x|\le 2} \int_0^1 (t^{\alpha p - n} \mu(B(x,t)))^{\frac{1}{p-1}} \frac{dt}{t} \, d\mu^F(x) \le c \cdot \mu(F)$$

$d > n - \alpha p$, $\alpha p < \lambda$. Hence

$$\mu(B \cap E_t)^{1/p} \le \frac{c}{t} \|f\|_{L^{p,\lambda}(B)},$$

which yields (a) for $1 \le s < p$. $\square$
Proof of (b): Writing

$$I_\alpha f(x) \int_0^\infty t^{\alpha-n} \int_{|x-y|<t} f(y) dy \frac{dt}{t}$$

we estimate as follows:

$$\int_0^1 \cdots \le \int_0^1 t^{\alpha-n} \|f\|_{L^{p,\lambda}(B)} \, t^{n-n/p} \, t^{(n-\lambda)/p} \frac{dt}{t} \le c \|f\|_{L^{p,\lambda}(B)}$$

when $\alpha p > \lambda$, and

$$\int_1^\infty \cdots \le \int_1^\infty t^{\alpha-n} \|f\|_{L^p(B)} t^{n-n/p} \, t^{(n-\lambda)/p} \frac{dt}{t} \le c \|f\|_{L^{p,\lambda}(B)},$$

when $\alpha p < n$. $\square$
Proof of (10.6): Using Hölders inequality, we can write

$$I_\alpha f(x) \le (I_{\alpha_1} f(x))^{1/q} \, (I_{\alpha_2} f(x))^{1/q'}$$

with $\alpha = \dfrac{\alpha_1}{q} + \dfrac{\alpha_2}{q'}$, $q' = q/(q-1)$, $q > 1$. And upon taking $\alpha_1 = \dfrac{n-d}{p} + \epsilon$, $\alpha_2 = \frac{\lambda}{p} + \epsilon$; $\epsilon > 0$ and small, we have

$$q = \frac{d + \lambda - n}{\lambda - \alpha p + \epsilon p}.$$

But then

$$(I_\alpha f(x))^{qs} \leq \left(\sup_B I_{\alpha_2} f \right)^{(q-1)s} (I_{\alpha_1} f(x))^s$$

Hence

$$\int_B (I_\alpha f(x))^{qs} \, d\mu(x) \leq c \, \|f\|_{L^{p,\lambda}(B)}^{(q-1)s} \, \|f\|_{L^{p,\lambda}(B)}^s$$

from Lemma 10.3; $qs < \dfrac{dp - p(n-\lambda)}{\lambda - \alpha p}$. □

Note: When $\lambda = n$, $q_0 < dp/(n - \alpha p)$ the correct limiting exponent of Theorem 10.1 ; $L^{p,n} = L^p$.

It should also be noted that Theorem 10.2 with exponent $q_0 < [dp - (n - \lambda)p]/(\lambda - \alpha p)$, $d > n - \alpha p, \alpha p < \lambda$, also follows from the weak-type result of [A4, Theorem 5.1]. Furthermore, with Theorem 10.2 a "gap" is created in the embedding.

$$I_\alpha : L^{p,\lambda}_{loc} \to L^q_{loc}(\mu), \tag{10.7}$$

$\mu = d$-measure, and $\frac{dp-(n-\lambda)p}{\lambda-\alpha p} \leq q \leq \frac{dp}{\lambda-\alpha p}$ since (10.7) is clearly false for $q > \frac{dp}{\lambda-\alpha p}$ via the function $f_0(y) = |y|^{-\lambda/p} \in L^{p,\lambda}_{loc}$ because $I_\alpha f_0(x) \sim |x|^{\alpha-\lambda/p}$ near zero. Also, this gap is closely related to finding lower bounds for $C_\alpha(\cdot\,; L^{p,\lambda})$ in terms of $\Lambda^d(\cdot), \lambda - \alpha p < d < n - \alpha p$.

Several attempts have failed to close this gap for arbitrary $f \in L^{p,\lambda}$ in a satisfactory manner; see [AX7]. So the question is: what replaces (10.7) for q in this gap?

Another observation with regard to (10.7) is that the vector functions $u(x) = \frac{x}{|x|^{\lambda/p}}$, $1 < p < \lambda$, are solutions to PDE - that of Koshelev; see Section 16.1 below. Here $|Du| \in L^{p,\lambda}_{loc}$ and $|u(x)| \in L^q_{loc}(\mu_d)$ for all $q < \frac{dp}{\lambda-p}$. So there seem to be "good" and "bad" Morrey-Sobolev functions related to (10.7).

10.3 An Improved Trace Result

One can slightly improve Theorem 10.2. Quoting from [AX7], we have

Theorem 10.4. For $\alpha p < \lambda$, $d > n - \alpha p, 1 < \gamma \leq q = \frac{dp-(n-\lambda)p}{\lambda-\alpha p}$, there is a constant $c > 0$ such that

$$\sup_f \int_B \left(\frac{I_\alpha f}{\|f\|_{L^{p,\lambda}}} \right)^q \left[\log \left(1 + \frac{I_\alpha f}{\|f\|_{L^{p,\lambda}}} \right) \right]^{-\gamma} d\mu < \infty. \tag{10.8}$$

and

Theorem 10.5. For $\alpha p = \lambda$, $d > n - \alpha p$,

$$\sup_f \int_B \exp\left(\frac{c\, I_\alpha f}{||f||_{L^{p,\lambda}}}\right) d\mu < \infty. \tag{10.9}$$

In both theorems, we assume $\mu = d$-measure, (10.2).

10.4 Notes

10.4.1. While on the subject of traces of Morrey potentials, it is interesting to point out that one can get a weighted embedding result on hyperplane in $\mathbb{R}^n$ - where μ is then the d-dimensional Lebesgue measure on the hyperplane on $\mathbb{R}^d \subset \mathbb{R}^n$. Further results of this type can be found in [AX7].

Theorem 10.6. For the weight $w(y) = |\hat{y}|^{\lambda - n + \epsilon}$, $\lambda \geq d$, $y = (\bar{y}, \hat{y}) \in \mathbb{R}^d \times \mathbb{R}^{n-d}$, there is a constant $c > 0$ such that

$$\left(\int_B |I_\alpha f|^q \, d\mu\right)^{1/q} \leq c\, ||f||_{L^p_w(B)} \tag{10.10}$$

for $q < \frac{dp}{\lambda - \alpha p}$, $\alpha p < \lambda$, and μ a d-measure (10.2) supported on $\mathbb{R}^d$; $d > \lambda - \alpha p$, $d \leq \lambda < n$.

Proof. (outline): we use duality, noting from [AX6] that the corresponding weighted Wolff potential is

$$W_{\alpha,p}^{\mu,w}(x) = \int_0^\infty \left(\frac{t^{\alpha p}\mu(B(x,t))}{w(B(x,t))}\right)^{\frac{1}{p-1}} \frac{dt}{t} \tag{10.11}$$

and for the weighted $w(y) = |\hat{y}|^{\lambda - n + \epsilon}$, we have

$$\fint_{B((\bar{x},0),t)} w(y)\, dy \sim t^{\lambda - n + \epsilon}.$$

But then (10.11) does not exceed

$$\int_0^\delta \left(\frac{t^{\alpha p + d}}{t^{\lambda + \epsilon}}\right)^{\frac{1}{p-1}} \frac{dt}{t} + ||\mu||^{\frac{1}{p-1}} \int_\delta^\infty \left(\frac{t^{\alpha p}}{t^{\lambda + \epsilon}}\right) \frac{dt}{t} \leq c\,||\mu||^{\frac{1}{p-1} + \frac{\alpha p - \lambda - \epsilon}{(p-1)d}}$$

upon choosing $\delta = ||\mu||^{1/d}$; $||\mu|| = \mu(\text{ supp }\mu)$. Thus it follows for $E_t = [I_\alpha f > t]$,

$$\mu(E_t) \leq c \left(\frac{||f||_{L^p(N)}}{t}\right)^q, \qquad q = \frac{dp}{\lambda - \alpha p + \epsilon}, \quad \epsilon > 0.$$

This is very interesting light of (10.7).   □

10.4.2. Notice that any CSI result implies a trace theorem; e.g., if $\mu(K) \leq c\, C_{\alpha,p}(K)$ for all compact sets K, $\mu =$ Borel measure, then the CSI

$$\int (I_\alpha f)^p \, dC_{\alpha,p} \leq c\, \|f\|_{L^p}^p \qquad (10.12)$$

implies

$$\int (I_\alpha f)^p \, d\mu \leq c\, \|f\|_{L^p}^p \qquad (10.13)$$

($f \geq 0$) a result similar to those above. The advantage that our theorems have over (10.12)–(10.13) is that we only assume our condition on μ for all balls - not for all compact sets!

Here is a short summary of the Maz'ya CSI type results in the literature - I apologize if I miss someone.

Some CSI/ Trace Time Lines

1964	V.G. Maz'ya [Ma1], CSI (10.12) for $\alpha = 1, 2$.
1971	D.R. Adams [A1], Traces of L^p potentials: Theorem 10.1.
1973	_____ [A2, A3], more trace results for potentials of L^p.
1973	_____ [A11], CSI (10.12) for $\alpha =$ positive integer.
1977	V. G. Maz'ya [Ma2], CSI for fractional Besov potentials.
1979	B. Dahlberg [Da], CSI (10.12) for fractional potentials.
1979	K. Hansson [H], CSI (10.12) for generalized potentials.
2003	D.R. Adams - J. Xiao [AX1]: CSI for Besov capacity.
2004	DR Adams - J, Xiao [AX2]: Failure of CSI for $X = L^{p,\lambda}$.

CSI = Capacity Strong-type Inequality

$$\int |I_\alpha f|^p \, dC \leq c\, \|f\|_X^p$$

where C is the capacity associated with Riesz potentials of functions in X. Generalized potentials include things like $k * f$, where $k = k(|x|)$ decreasing to zero as $|x| \to \infty$; or even some signed kernels, perhaps.

Chapter 11
Interpolation of Morrey Spaces

11.1 Stampacchia-Peetre interpolation; Interpolation via the new duality

Now we turn our attention to interpolation of linear operators on Morrey spaces, say

$$T : L^{p,\lambda} \longrightarrow L^{q,\mu}$$

for various p, q, λ, and μ; $1 < p, q < \infty$, $0 < \lambda$, $\mu < n$.

First, with regard to the history of such attempts. It must surely start with Stampacchia [S] in 1965 when he showed that interpolation works quite easily when the Morrey Spaces lie only in the range of the operator. In fact, if

$$T : L^{q_i} \longrightarrow L^{p_i,\lambda_i}, \ i = 0, 1$$

then

$$T : L^{q_\theta} \longrightarrow L^{p_\theta,\lambda_\theta}$$

where

$$\frac{1}{q_\theta} = \frac{1-\theta}{q_0} + \frac{\theta}{q_1}, \ \frac{1}{p_\theta} = \frac{1-\theta}{p_0} + \frac{\theta}{p_1} \qquad (11.1)$$

and

$$\frac{\lambda_\theta}{p_\theta} = (1-\theta)\frac{\lambda_0}{p_0} + \theta\frac{\lambda_1}{p_1}, \qquad (11.2)$$

and $0 < \theta < 1$.

© Springer International Publishing Switzerland 2015
D. Adams, *Morrey Spaces*, Applied and Numerical Harmonic Analysis,
DOI 10.1007/978-3-319-26681-7_11

Proof. The hypotheses imply

$$\sup_{x,r>0} \left(r^{\lambda_i - n} \int_{B(x,r)} |Tf|^{p_i} \, dy \right)^{1/p_i} \leq A_i \left(\int |f|^{q_i} \, dy \right)^{1/q_i},$$

for $i = 0, 1$. Hence for any fixed ball $B(x, r)$,

$$\int_{B(x,r)} |Tf|^{p_i} \, dy \leq A_i^{p_i} \, \|f\|_{L^{q_i}}^{p_i} \cdot r^{n - \lambda_i}.$$

So now by standard interpolation between Lebesgue spaces, one has

$$\left(\int_{B(x,r)} |Tf|^{p_\theta} \, dy \right)^{1/p_\theta} \leq M_\theta \, \|f\|_{L^{q_\theta}}$$

with $M_i = A_i \, r^{(n - \lambda_i)/p_i}, \quad i = 0, 1.$
Hence $M_\theta = M_0^{1-\theta} M_1^{\theta} = A_0^{1-\theta} A_1^{\theta} r^{(n - \lambda_\theta)/p_\theta}.$
Which gives

$$\|Tf\|_{L^{p_\theta, \lambda_\theta}} \leq M_\theta \, \|f\|_{L^{q_\theta}}.$$

$\square$

But this argument does not work for

$$T : L^{p_i, \lambda_i} \longrightarrow L^{q_i}, \ i = 0, 1.$$

So we will concentrate on this case now. And, of course, to prove this, it is now natural to use the duality that we have established in Chapter 5. Thus let $T^* =$ adjoint of T and suppose that

$$T^* : L^{q_i'} \longrightarrow H^{p_i', \lambda_i}, \ i = 0, 1.$$

The hypothesis implies that there are $\omega_i \in A_1$ such that

$$\int \omega_i \, d\Lambda^{n - \lambda_i} \leq 1$$

and

$$\left(\int |T^*f|^{p_i'} \omega_i^{1 - p_i'} \, dy \right)^{1/p_i'} \leq A_i \left(\int |f|^{q_i'} \, dy \right)^{1/q_i'}.$$

We now write

$$\omega_i^{1 - p_i'} = (\omega_i^{-1/p_i})^{p_i'}. \tag{11.3}$$

This then allows us to apply "Stein's Interpolation with change of measure" (see [BS]) to get

$$\left(\int |T^*f|^{p'_\theta} \, \omega_\theta^{1-p'_\theta} \, dy \right)^{1/p'_\theta} \le A_\theta \left(\int |f|^{q'_\theta} \, dy \right)^{1/q'_\theta}$$

where

$$\omega_\theta = \omega_0^{\frac{p_\theta}{p_0}(1-\theta)} \cdot \omega_1^{\frac{p_\theta}{p_1}\theta}. \tag{11.4}$$

To make our conclusion and then by duality our interpolation result, we must now show that

$$\omega_\theta \in A_1 \text{ and } \int \omega_\theta \, d\Lambda^{n-\lambda_\theta} \le \text{constant}.$$

We first note that this is now very easy in the case where $\lambda_0 = \lambda_1 = \lambda$, i.e., in the known case in the literature [Y², Y²Z], because (11.4) implies that $\omega_\theta \in A_1$ and

$$\int \omega_\theta \, d\Lambda^{n-\lambda} \le c \left(\int \omega_0 \, d\Lambda^{n-\lambda} \right)^{\frac{1-\theta}{p_0} p_\theta} \left(\int \omega_1 d\Lambda^{n-\lambda} \right)^{\frac{\theta}{p_1} p_\theta}$$

by the quasi Holder (via $\tilde{\Lambda}_0^{n-\lambda}$), i.e., $\frac{p_\theta}{p_0}(1-\theta) + \frac{p_\theta}{p_1}\theta = 1$.

Thus we get

$$\|T^*f\|_{H^{p'_\theta,\lambda}} \le A_\theta \, \|f\|_{L^{q'_\theta}}$$

and then for our result

$$T : L^{p_\theta,\lambda} \to L^{q_\theta}$$

for the case $\lambda_0 = \lambda_1 = \lambda$.

However, we are trying for more. And to do this we now invoke the atomic decompositions of [AX2], we have

Lemma 11.1. $\omega \in L^1(\Lambda^d)$ iff $\omega = \sum_k c_k \, a_k$, where $\{c_k\} \in l^1$ and the a_k are (∞, d)- atoms, i.e.

(i) supp $a_k \subset$ cube Q_k

and

(ii) $\|a_k\|_{L^\infty(Q_k)} \le |Q_k|^{-d/n}$.

And the norm of $L^1(\Lambda^d)$ (actually a quasi-norm for we must go through $\tilde{\Lambda}_0^d$ - see Chapter 3), is equivalent to

$$\inf \sum_k |c_k|$$

with the infimum over all such representations.

With this lemma, we argue as follows: write

$$\omega_i = \sum_k C_k^{(i)} a_k^{(i)}, \quad i = 0, 1, \text{ with}$$

$$\sum_k |C_k^{(i)}| \sim \int \omega_i \, d\Lambda^{n-\lambda_i}.$$

Then set

$$C_k = |C_k^{(0)}|^{1-\Theta} |C_k^{(1)}|^{\Theta},$$

with

$$\Theta = \frac{p_\theta}{p_1} \theta \quad \text{and} \quad 1 - \Theta = 1 - \frac{p_\theta}{p_1}\theta = \frac{p_\theta}{p_0}(1 - \theta).$$

Then

$$\sum_k C_k \leq \left(\sum_k |C_k^{(0)}| \right)^{1-\Theta} \left(\sum_k |C_k^{(1)}| \right)^{\Theta}$$

$$\leq A \left(\int \omega_0 \, d\Lambda^{n-\lambda_0} \right)^{1-\Theta} \left(\int \omega_1 \, d\Lambda^{n-\lambda_1} \right)^{\Theta}$$

$$\leq A = \text{absolute constant.}$$

Thus setting $a_k^{(\theta)} = |a_k^{(0)}|^{1-\Theta} \cdot |a_k^{(1)}|^{\Theta}$, we see that it is indeed an $(\infty, n - \lambda_\theta)$ atom. This then is our argument for the

Theorem 11.2. If the linear operator T satisfies

$$T : L^{p_i, \lambda_i} \longrightarrow L^{q_i}, \quad i = 0, 1$$

with $1 < p_i < \infty$ and $0 < \lambda_i < n$, then there is a constant such that

$$T : L^{p_\theta, \lambda_\theta} \longrightarrow L^{q_\theta}$$

with $p_\theta, q_\theta, \lambda_\theta$ given by (11.1) and (11.2).

Proof. Proof of Lemma 11.1

On one hand, if $f = \sum_k c_k \, a_k$, then by the quasi-sublinearity

$$\|f\|_{L^1(\Lambda^d)} \leq \int \sum_k |c_k| \, |a_k| \, d\Lambda^d$$

$$\leq c \sum_k |c_k| \int |a_k| \, d\Lambda^d$$

$$\leq c \, \|\{c_k\}\|_{l^1}.$$

Conversely, suppose $\|f\|_{L^1(\Lambda^d)} < \infty$, then using the construction given in [AX2], we can write

$$f = \sum_{j,k} c_{j,k} a_{j,k}$$

where

$$c_{j,k} = l(Q_{j,k})^d \cdot 2^{k+1}$$
$$a_{j,k}(x) = f(x) \cdot X_{\Delta_{j,k}(x)} \cdot l(Q_{j,k})^{-d} \cdot 2^{-(k+1)}$$

$l(Q) =$ edge length of the cube Q. Then $|f(x)| \leq 2^{k+1}$ for $x \in \Delta_{j,k}$ and $\{c_{j,k}\} \in l^1$ since

$$\|\{c_{j,k}\}\|_{l^1} \leq A \sum l(Q_{j,k})^d \, 2^{k+1} \leq A \, \|f\|_{L^1\Lambda^d}.$$

The $Q_{j,k}$ and $\Delta_{j,k}$ are selectively chosen dyadic cubes; see[AX2].   □

Finally one can argue for

$$T : L^{p,\lambda} \longrightarrow L^{q,\mu}$$

by putting together our argument with that of Stampacchia.

11.2   Counterexamples to interpolation with Morrey Spaces in the domain of the operator

But one should note that there are some counterexamples. A close examination of these examples shows in fact that the Campanato spaces are not stable under interpolation. In [BRV], the authors give an example (in one dimension) where

$0 < \lambda_0 < \lambda_1 = 1 = n$, and then show that interpolation does not achieve a Morrey Space estimate for the intermediate case of $\lambda_\theta : 0 < \lambda_0 < \lambda_\theta < \lambda_1 = 1$. Also, from [SZ] one gets

$$T : C^\alpha \longrightarrow C^\alpha$$
$$T : L^2 \longrightarrow L^2$$

but $T(f) \notin L^q$ for any $q > 2$. One would hope for some Morrey cases as intermediate situations. Also, we note from the proof given for Theorem 11.2, that it is really necessary to have $\lambda \in (0, n)$ for $\int \omega \, d\Lambda^{n-\lambda}$ doesn't make sense when $\lambda = n$.

11.3   Integrability of Morrey Potentials

As an application of Theorem 11.2, we get the following integrability result for Morrey potentials: $I_\alpha f, f \in L^{p,\lambda}, \alpha p < \lambda < n$. Indeed, Lemma 9.1 yields

$$I_\alpha : L^{p,\lambda_0} \longrightarrow \text{BMO} \subset L^{q_0}_{loc}$$

for any $q_0 < \infty$ when $\lambda_0 = \alpha p < n$, $p > 1$. And Theorem 7.1(i)

$$I_\alpha : L^{p,\lambda_1} \longrightarrow L^{q_1}_{loc}$$

for $q_1 = \lambda_1 p/(\lambda_1 - \alpha p)$, $\alpha p < \lambda_1 < n$, $p > 1$. Hence by interpolation

$$I_\alpha : L^{p,\lambda} \longrightarrow L^q_{loc}$$

for any $q < np/(\lambda - \alpha p)$, $\alpha p = \lambda_0 < \lambda < \lambda_1 < n$.

And notice, $I_\alpha f_0 \in L^q_{loc}$ for $f_0(y) = |y|^{-\lambda/p}$, $y \in \mathbb{R}^n$.

This all seems to work out here since λ_0 and λ_1 both less that $n=$ dimension of the underlying space, hence no L^p spaces are included in the class of functions being interpolated (see [BRV] and [LR]).

Chapter 12
Commutators of Morrey Potentials

12.1 Some history for the operators $[b, T]$ and $[b, I_\alpha]$

In this chapter, we will briefly study the commutator operators (by multiplication) on the Morrey space of potentials. So given an operator T (from say L^p into L^q, some p and q), its commutator by multiplication by function $b(x)$ is given by

$$[b, T]f = b(x) \cdot Tf(x) - T(b \cdot f)(x)$$

provided all the terms make sense; e.g., when $b \in L^\infty(\mathbb{R}^n)$. One of the first results on such operators, say on the Lebesgue classes, was due to Coifman-Rochberg-Weiss [CRW] 1976, though their investigations were very much influenced by A.P. Calderon and his commutator operators on Lipschitz curves; see [To] for some discussion of this. But before we get to the results, here are some time lines:

Some Commutator Time Lines

($K = CZ$ operator, $I_\alpha =$ Riesz potential operator.)

1976 Coifman, R., Rochberg, R., G. Weiss. [CRW]:
   $[b, K] : L^p \to L^p$ iff $b \in$ BMO $(1 < p < \infty)$
1978 Uchiyama, A. [U]:
   $[b, K] : L^p \to L^p$ is compact iff $b \in$ VMO $(1 < p < \infty)$
1982 Chanillo, S. [Ch]:
   $[b, I_\alpha] : L^p \to L^{p^*}$ iff $b \in$ BMO $(p^* = np/(n - \alpha p))$
1991 Fazio, G., Ragusa, M. [FR]:
   $[b, K] : L^{p,\lambda} \to L^{p,\lambda}$ iff $b \in$ BMO
   $[b, I_\alpha] : L^{p,\lambda} \to L^{\tilde{p},\lambda}$ if $b \in$ BMO, $\tilde{p} = \left(\dfrac{\lambda p}{\lambda - \alpha p} \right)$

© Springer International Publishing Switzerland 2015
D. Adams, *Morrey Spaces*, Applied and Numerical Harmonic Analysis,
DOI 10.1007/978-3-319-26681-7_12

1997   Ding, Y [D]::

      $[b, I_\alpha] : L^{p,\lambda} \to L^{\tilde{p},\lambda}$ iff $b \in$ BMO

2003   Komori, Y., Mizahara, T. [KM]

      Independent proof of [D]

2009   Chen, Y., Ding, Y., Wang, X. [CDW]:

      $[b, I_\alpha] : L^{p,\lambda} \to L^{\tilde{p},\lambda}$ is compact iff $b \in$ VMO

2011   Adams, DR., Xiao, J, [AX4]:

      (i) The following are equivalent

          (a) $b \in$ BMO

          (b) $[b, I_\alpha] : H^{p,\lambda} \to H^{\tilde{p},\lambda}$

          (c) $[b, I_\alpha] : L^{p,\lambda} \to H^{\tilde{p}',\lambda}$

      (ii) If $|\nabla b| \in L^n$ and $\mu(B(x,r)) \leq C\, r^d$, all x and $r > 0$, then

          $[b, I_\alpha] : L^{p,\lambda} \to L^q(\mu), \; q < \frac{dp-(n-\lambda)p}{\lambda-\alpha p}$.

For the proofs of these results, see esp [To], upto 2009, for they all now come from a key lemma - and according to [To], this approach is actually due to J. O. Stromberg. And this is our first result; its modification (Lemma 12.2) is due to [FR].

Lemma 12.1. For $T = K$, a Calderon-Zygmund operator, one has

$$([b,K]f)^\#(x) \leq c \, ||b||_{\text{BMO}} \left[(M_0 \, |Kf|^r)^{1/r}(x) + (M_0 \, |f|^s)^{1/s})(x) \right], \qquad (12.1)$$

with $1 < r, s < \infty$.

Lemma 12.2. For $T = I_\alpha$, the Riesz potential operator, one has

$$([b,I_\alpha]f)^\#(x) \leq c \, ||b||_{\text{BMO}} \left[(M_0 \, |I_\alpha f|^r)^{1/r}(x) + (M_{\alpha s} \, f^s)^{1/s})(x) \right], \qquad (12.2)$$

with $1 < r, s < \infty$.

Here for convenience, we have assumed $f \geq 0$. The proof of Lemma 12.1 can be found [To] page 418. The proof of Lemma 12.2 is similar.

12.2   Commutators: $b \in$ BMO

In this section, we will give the arguments for (a) $\Longrightarrow$ (b) "from [AX4]; see 2011 on page 69." (For (b) $\Longrightarrow$ (c), we simply note

$$\int [b, I_\alpha] f \cdot g \, dx = - \int [b, I_\alpha] g \cdot f \, dx$$

use duality and (a) $\implies$ (b). We refer the reader to the literature for (c) $\implies$ (a); say [D] or [CDW]). So using Lemma 12.1 and the basic inequality of Fefferman-Stein [St1], [ST] we can write

$$\left| \int [b, I_\alpha] f \cdot g \, dx \right| \leq \int ([b, I_\alpha] f)^\# \cdot M_0 \, g \, dx.$$

Hence we just need to estimate

$$\int (M_0 |I_\alpha f|^r)^{1/r} M_0 \, g \, dx = P_1$$

and

$$\int (M_{\alpha s} f^s)^{1/s} M_0 \, g \, dx = P_2.$$

Here we choose $1 < r, s < p < \lambda/\alpha$, $f \geq 0$.

$$P_1 \leq \inf_\omega \left(\int [M_0 (I_\alpha f)^r]^{q/r} \, \omega^{1-q} \, dx \right)^{1/q} \cdot ||M_0 \, g||_{L^{q', \lambda}}$$

with $\omega \in A_1$ and $\int \omega \, d\Lambda^{n-\lambda} \leq 1$ as per $H^{q, \lambda}$. Hence it follows that

$$P_1 \leq c \, ||f||_{H^{p, \lambda}} \, ||g||_{L^{q', \lambda}},$$

since $\omega^{1-q} \in A_{q/r}$ via the A_p - weight theory; see [To]. Also, $q = \lambda p / (\lambda - \alpha p)$. Next,

$$P_2 \leq \left(\inf_\omega \int (M_{\alpha s} f^s)^{q/s} \, \omega^{1-q} \, dx \right)^{1/q} \cdot ||M_0 \, g||_{L^{q', \lambda}}.$$

And again using the full strength of the A_p - weight theory, we can write

$$\omega^{1-q} = W^{1-\frac{q}{s}}, \qquad W = \omega^{(q-1)/(\frac{q}{s}-1)} \in A_1.$$

Thus,

$$P_2 \leq ||M_{\alpha s} f^s||_{H^{q/s, \lambda}}^{1/s} \cdot ||g||_{L^{q', \lambda}}$$

$$\leq c \, ||f^s||_{H^{p/s, \lambda}} \cdot ||g||_{L^{q', \lambda}}$$

because $M_{\alpha s} f^s \leq c \, I_{\alpha s} f^s$. Hence

$$P_2 \leq c \, ||f||_{H^{p, \lambda}} \cdot ||g||_{L^{q', \lambda}}. \qquad \qquad \square$$

We now mention a borderline case.

Theorem 12.3. For $1 < p = \lambda/\alpha$, $0 < \alpha, \lambda < n$, if $b \in$ BMO, then

$$\sup_{\|f\|_{L^{p,\lambda}} \leq 1} \int_{B(0,1)} \exp \left| \frac{c[b, I_\alpha]f}{\|b\|_{\text{BMO}}} \right| dx < \infty. \tag{12.3}$$

See [AX4].

12.3 Traces of Morrey commutators, $|\nabla b| \in L^n$

We now turn our attention to the main new result of [AX4], namely a trace result for Morrey commutators. Clearly, since $[b, I_\alpha]f = b \cdot I_\alpha f - I_\alpha(bf)$, one cannot expect a finite trace on any lower dimensional subset of $\mathbb{R}^n$ because if $b(x)$ is merely BMO, it is only defined $\mathscr{L}^n$ -a.e. Although, $I_\alpha f$ and $I_\alpha f(bf)$ do have a trace on lower dimensional sets, it is clear that to expect a finite trace from the commutator, we must require that b have additional smoothness. Our assumption will be that $|\nabla b| \in L^n(\mathbb{R}^n)$. Notice that this assumption implies that $b \in$ BMO by the standard Sobolev estimates.

Theorem 12.4. Let $|\nabla b| \in L^n$ and $0 < \alpha < n$, $1 < p < \lambda/\alpha$, $0 < \lambda < n$, then for μ a Borel measure satisfying

$$\mu(B(x, r)) \leq C \, r^d$$

for all x and $r > 0$, with $d > n - \alpha p$,

$$\sup_{\|f\|_{L^{p,\lambda}} \leq 1} \int_{B(0,1)} |[b, I_\alpha]f|^q \, d\mu < \infty \tag{12.4}$$

when either $q < \dfrac{dp - (n - \lambda)p}{\lambda - \alpha p}$ for $\lambda < n$, or $q \leq \frac{dp}{n - \alpha p}$ for $\lambda = n$.

Proof.   Case 1. $\alpha = 1$

$$\nabla[b, I_1]f(x) = [b, \nabla I_1]f(x) + \nabla b \cdot I_1 f(x)$$

with ∇I_1 a vector of CZ singular integrals. Thus

$$
\begin{aligned}
\|\nabla[b, I_1]f\|_{L^{p,\lambda}} &\leq \|[b, \nabla I_1]f\|_{L^{p,\lambda}} + \| |\nabla b| \cdot I_1 f\|_{L^{p,\lambda}} \\
&\leq c \, \|b\|_{\text{BMO}} \|f\|_{L^{p,\lambda}} + \|\nabla b\|_{L^{\lambda,\lambda}} \|f\|_{L^{p,\lambda}} \\
&\leq c \, (\|b\|_{\text{BMO}} + \|\nabla b\|_{L^n}) \cdot \|f\|_{L^{p,\lambda}} \\
&\leq c \, \|\nabla b\|_{L^n} \|f\|_{L^{p,\lambda}}.
\end{aligned}
$$

Now set $h = |\nabla[b, I_1]f|$, then

$$|[b, I_1]f| \le c \cdot I_1 h, \ h \in L^{p,\lambda},$$

Since for smooth φ with compact support, one has

$$\varphi(x) = c \sum_{i=1}^{n} \int \frac{x_i - y_i}{|x - y|^n} \cdot \frac{\partial}{\partial x_i} \varphi(y) \, dy$$

via Fourier transform consideration. Thus (12.4) follows from Theorems 10.2 and 10.1.

Case 2. $\alpha = 2$. Now write

$$I_\theta([b, I_1]f) = [b, I_{1+\theta}]f - [b, I_\theta](I_1 f). \tag{12.5}$$

Here we have used the semi-group property for correctly normalized Riesz potential operators: $I_\alpha I_\beta = I_{\alpha+\beta}$. Thus when $\theta = 1$, (12.5) is just

$$|[b, I_2]f| \le c [I_1 g + I_2 h]$$

where $g \in L^{\tilde{p},\lambda}$, $\tilde{p} = \lambda p/(\lambda - p)$, and $h \in L^{p,\lambda}$. Hence

$$\left(\int_{B(0,1)} |I_1 g|^q \, d\mu \right)^{1/q} \le c \, \|f\|_{L^{p,\lambda}}$$

for $q < (dp - (n - \lambda)p)/(\lambda - 2p)$ and

$$\left(\int_{B(0,1)} |I_2 h|^q \, d\mu \right)^{1/q} \le c \, \|f\|_{L^{p,\lambda}},$$

for the same exponent q, by Theorems 10.1 and 10.2.

Case 3 $\alpha = m \in \mathbb{N}$. The idea now is to proceed by induction on (12.5).

$$|[b, I_{m+1}]f| \le c \ (|[b, I_m](I_1 f)| + |I_m[b, I_1]f|)$$

$$\le c \sum_{j=0}^{m} I_{m+1-j} f_j$$

with the usual Morrey estimates, $q < [dp - (n - \lambda)p]/(\lambda - (m+1)p)$, etc. Details are left to the reader.

Case 4. $m - 1 < \alpha < m$, $m \in \mathbb{N}$. For simplicity, we look only at the sub case: $0 < \alpha < 1$.

Here we define the vector valued operator

$$T_\alpha(b,f)(x) = \nabla(I_{1-\alpha}[b, I_\alpha]f)(x) \tag{12.6}$$

for $0 < \alpha < 1$. We intend to show

$$||T_\alpha(b,f)||_{L^{p,\lambda}} \leq c\, ||\nabla b||_{L^n} ||f||_{L^{p,\lambda}} \tag{12.7}$$

for all $0 < \alpha < 1$; $1 < p < \infty$, $0 < \lambda < n$. And to see (12.7) we complexify α as $\zeta + i\eta$, $0 \leq \zeta \leq 1$, $\eta \in \mathbb{R}$, and use Hadamard's Three-lines Lemma; see [BS].
    Thus we write

$$H(\zeta + i\eta) = \int T_{\zeta+i\eta}(b,f) \cdot \bar{g}\, dx$$

where $\bar{g}$ is an n-vector of $H^{p',\lambda}$ functions.
    So H is a complex analytic function on the strip $0 < \zeta < 1$, $\eta \in \mathbb{R}$. And we intend to get estimates of the form

$$|H(0 + i\zeta)| \leq M_0 ||\nabla b||_{L^n} ||f||_{L^{p,\lambda}} \cdot ||\bar{g}||_{H^{p',\lambda}} \tag{12.8}$$

and

$$|H(1 + i\zeta)| \leq M_1 ||\nabla b||_{L^n} ||f||_{L^{p,\lambda}} \cdot ||\bar{g}||_{H^{p',\lambda}}.$$

It then follows that

$$|H(\alpha)| \leq M_\alpha ||\nabla b||_{L^n} ||f||_{L^{p,\lambda}} \cdot ||\bar{g}||_{H^{p',\lambda}}$$

which gives (12.7) by duality.

Notice that from this and the use of the Fourier transform, if $h = T_\alpha(b,f)$, then

$$[b, I_\alpha]f = \bar{R}I_\alpha\, h = I_\alpha\bar{R}\, h$$

So

$$|[b, I_\alpha]f| \leq c \cdot I_\alpha|\bar{R}\, h|$$

$\bar{R} = $ vector of n-Riesz transforms; $||\bar{R}\, h||_{L^{p,\lambda}} \leq c\, ||f||_{L^p}$. Thus we estimate

$$||T_{i\eta}(b,f)||_{L^{p,\lambda}} \leq ||\nabla I_{1+i\eta}[b, I_{i\eta}]f||_{L^{p,\lambda}}$$

$$\leq c\, ||f||_{L^{p,\lambda}} \tag{12.9}$$

Since $\nabla I_{1-i\eta}$ is a CZ operator as is $I_{i\eta}$. The constants in (12.9) are independent of η. Next,

$$T_{1+i\eta}(b,f) = \nabla(I_{-i\eta}(b, I_{1+i\eta}f) - I_{1+i\eta}(bf))$$
$$= I_{-i\eta}((\nabla b)I_{1+i\eta}f) + I_{-i\eta}[b, \bar{K}]f$$

$\bar{K} = \nabla|x|^{1+i\eta-n}$, again a CZ singular integral kernel.

Hence

$$||T_{1+i\eta}(b,f)||_{L^{p,\lambda}} \leq c \left(||b||_{\text{BMO}} ||f||_{L^{p,\lambda}} + ||\nabla b||_{L^n} ||f||_{L^{p,\lambda}} \right)$$

as before. This concludes our proof of case 4. The remaining cases can be handed by induction; see [AX4]. □

It is interesting here that $|\nabla b| \in L^n$ works just as well for α large or small. This suggests that especially for small α that $|\nabla b| \in L^n$ may not be necessary. Perhaps one can get by with $b = I_\epsilon h$, with

$$h \in L^{n/\epsilon}, \ 0 < \epsilon < 1$$

And if not, why not?

And finally, we mention a limiting case result: $\alpha p = \lambda$; see [AX4].

Theorem 12.5. Let $b = I_\alpha h$, $h \in L^{n/\alpha}$, $p > 1$, and μ a d-measure, $d > n - \alpha p > 0$. Then

$$\sup_{||f||_{L^{p,\lambda}} \leq 1} \int_{B(0,1)} \exp(c |[b, I_\alpha]f|^q) \, d\mu < \infty$$

for either $0 < \lambda < n$, $q < 1$ or $\lambda = n$, $q = p'$.

Chapter 13
Mock Morrey Spaces

13.1 Marcinkiewicz Spaces

The use of the word "mock" here is essentially in tribute to Ramanujan's use of the word when he refers to his "mock theta functions" - they have a close resemblance to the original (imitations), but are not quite the same. This is the case with the spaces of this chapter, and the term is to draw attention to this close resemblance. Actually, these spaces can also be described by the term "Marcinkiewicz Spaces," but in so doing, the underlying connection with Morrey's original tends to be lost.

A Mock Morrey Space for us, loosely speaking, will be any space, initially defined for arbitrary sets E (usually restricted to the class of all compact sets of $\mathbb{R}^n$), which becomes exactly a Morrey Space when the set E is replaced by a ball or a cube (with sides parallel to the axes). Of main interest here are those $f(y)$ that satisfy

$$\left(\sup_E S(E)^{(\lambda-n)/\sigma} \int_E |f(y)|^p \, dy \right)^{1/p} = \|f\|_{Q^{p,\lambda}(S)} < \infty. \tag{13.1}$$

where S is a non-negative set function on subsets of $\mathbb{R}^n$ such that $S(B(x,r)) \sim r^\sigma$, for all $r > 0$. Standard set functions used here may include Lebesgue measure $\mathscr{L}^n$, Hausdorff capacity Λ^d, or any of the Riesz L^p capacities $\dot{C}_{\alpha,p}$, or even $\dot{C}_\alpha(\cdot \, ; \, L^{p,\lambda})$. When

$$S = \dot{C}_{\alpha,p}(\cdot), \quad \sigma = n - \alpha p$$

the $Q's$ coincide with a space of Sobolev multipliers

$$\gamma \in M[\dot{L}^{\alpha,p} \longrightarrow L^p];$$

see [MS1] and [MS2].

© Springer International Publishing Switzerland 2015
D. Adams, *Morrey Spaces*, Applied and Numerical Harmonic Analysis,
DOI 10.1007/978-3-319-26681-7_13

Notice, that one easily has the inclusions

$$L^{p,\lambda} \supset Q^{p,\lambda}(\mathscr{L}^n) \supset Q^{p,\lambda}(\dot{C}_{\alpha,p}) \supset Q^{p,\lambda}(\dot{C}_{\beta,p}),$$

for $0 < \alpha < \beta$, Also, note that the function

$$|y|^{-\lambda/p} \in Q^{p,\lambda}(\mathscr{L}^n).$$

13.2   Conti's Theorem

Our main result here is an extension of a result due to Conti [Co].

Theorem 13.1. If $f \in Q^{p,\lambda}(\dot{C}_{\alpha,p})$, then there is a constant $A > 0$ such that

$$\dot{C}_{\alpha,p}([|I_\alpha f| > t]) \leq A \, t^{-q_0} \, ||f||^{q_0}_{Q^{p,\lambda}(\dot{C}_{\alpha,p})} \tag{13.2}$$

for $q_0 = (n - \alpha p)p/(\lambda - \alpha p)$, $\alpha p < \lambda$, $p > 1$. And for $\alpha p = \lambda$

$$C_{\alpha,p}([|G_\alpha f| > t]) \leq A \, e^{-a \, t/||f||}, \tag{13.3}$$

for some constants a and A independent of f; $||f||$ is exactly the $Q^{p,\lambda}(C_{\alpha,p})$ norm.

Proof. The proof of this theorem requires

(a) The classical operators of Harmonic Analysis (i.e., M_0 and $K = CZ$ operator) are bounded on the $Q^{p,\lambda}$ spaces. This is a result originally due to Verbitsky; see [MS1].
(b) an iteration scheme

$$\phi(2^k) \leq A \, 2^{-kp} \, \phi(2^{k-1})^\delta \tag{13.4}$$

for $\phi(t)$ a decreasing function of $t > 0$ and some $\delta \in (0, 1)$. Or

$$\phi(h_s) \leq \frac{1}{e}\phi(h_{s-1}) \tag{13.5}$$

with $h_s = k_0 + s(Ae)^{1/p}$, $s > 0$.

We give only an outline of the argument for it seems that these spaces are of only marginal interest to those who find merit with the Morrey Scale. To see (a), we must have an important result of Verbitsky:

Lemma 13.2. The Wolff potentials $W^\mu_{\alpha,p}$ for a $C_{\alpha,p}$ capacity measure μ satisfies

$$[\dot{W}^\mu_{\alpha,p}]^\beta \in A_1$$

the Muckenhoupt weight class; $0 < \beta < (p-1)n/(n - \alpha p)$.

Thus if $\dot{W}_{\alpha,p}^{\mu} \geq 1$ on E, then

$$\int_E (M_0 f)^p \, dx \leq \int (M_0 f)^p (\dot{W}_{\alpha,p}^{\mu})^{\beta} \, dx \leq A \int |f|^p (\dot{W}_{\alpha,p}^{\mu})^{\beta} \, dx$$

$$= A \int_0^{\infty} \int_{S_t} |f|^p \, dx \, t^{\beta-1} \, dt$$

$$\leq A \int_0^M \dot{C}_{\alpha,p}(S_t) \, t^{\beta-1} \, dt \cdot \|f\|_Q^p$$

$$\leq A \int_0^M t^{-(p-1)\delta} \, t^{\beta-1} \, dt \cdot \|f\|_Q^p \cdot \dot{C}_{\alpha,p}(E)^{\delta}, \qquad (13.6)$$

$\delta = (n-\lambda)/(n-\alpha p)$, and where M is the upper bound for $\dot{W}_{\alpha,p}^{\mu}$; see [AH]. Also, to get line (13.6), we use a standard estimate for $\dot{C}_{\alpha,p}(S_t)$, $S_t = [\dot{W}_{\alpha,p}^{\mu} > t]$; see [AH], page 164. Thus we get

$$\|M_0 f\|_{Q^{p,\lambda}(\dot{C}_{\alpha,p})} \leq A \, \|f\|_{Q^{p,\lambda}(\dot{C}_{\alpha,p})}.$$

A similar argument works for K, and this proves part (a).

For a proof Lemma 13.2, the reader is referred to [MS1], page 100, where the lemma is given for $U_{\alpha,p}^{\mu}(x) = I_{\alpha}(I_{\alpha}\mu)^{\frac{1}{p-1}}$, the non-linear Riesz potential of the measure μ. Replacing $U_{\alpha,p}^{\mu}$ by $W_{\alpha,p}^{\mu}$ works upon consulting [AH] and the relationship between these two non-linear potentials: for $p > 2 - \alpha/n$, $W_{\alpha,p}^{\mu} \sim U_{\alpha,p}^{\mu}$, but for $1 < p < 2 - \alpha/n$, this does not hold, only $W_{\alpha,p}^{\mu} \leq c \, U_{\alpha,p}^{\mu}$ holds in this case. But one need not do this. It is easy to see that $(W_{\alpha,p}^{\mu})^{\beta} \in A_1$ directly.

Recall: $\omega \in A_1$ iff $\fint_Q \omega \leq c \cdot \inf_Q \omega$.

To see (b), let us look at the case of $\alpha \in \mathbb{N}$ for simplicity; fractional values of α can be handled by the trick of Dahlberg [D] and will not be treated here; see [A10].

So let $H \in C^{\infty}(0, \infty)$ such that

$$H(t) = \begin{cases} 0, \, t \leq 1/2 \\ 1, \, t \geq 1. \end{cases}$$

Then as in [A11]

$$\int \left| D^{\alpha} H\left(\frac{I_{\alpha} f}{2^k}\right) \right|^p \, dx \leq A \int_{E_{k-1}} (M_0 f)^p \, dx \cdot 2^{-kp}$$

$$\leq A \int (M_0 f)^p \, (W_{\alpha,p}^{\mu})^{\beta} \, dx \cdot 2^{-kp}$$

with

$$E_{k-1} = [I_\alpha f > 2^{k-1}], \quad f \geq 0.$$

But then

$$\dot{C}_{\alpha,p}([I_\alpha f > 2^k]) \leq A \, \|f\|^p_{Q^{p,\lambda}(\dot{C}_{\alpha,p})} \cdot 2^{-kp} \cdot \dot{C}_{\alpha,p}([I_\alpha f > 2^{k+1}])^\delta \qquad (13.7)$$

with $\delta = (n-\lambda)/(n-\alpha p)$, $\alpha p < \lambda$. (13.7) is just (13.3).

Upon iterating (13.3), we get

$$\phi(t) \leq \frac{\phi(1) \, A^{1/(1-\delta)} \, 2^{p/(1-\delta)^2}}{t^{p/(1-\delta)}},$$

which gives (13.2). For (13.3), we get respectively,

$$\phi(t) \leq A \, e^{-t}$$

upon iteration of (13.5).   □

13.3   Notes

13.3.1 $Q^{p,\lambda}$ vs $L^{p,\lambda}$

One very clear difference between Morrey Spaces and Mock Morrey Spaces is that if

$$f \in Q^{p,\lambda}(\mathscr{L}^n)$$

then $f \in L^{(q,\lambda)}$, $q = np/\lambda > p$, the Lorentz space. Higher integrability does not occur in the Morrey case; i.e., if $f \in L^{p,\lambda}$, then one cannot generally conclude that $f \in L^q_{loc}$ for any $q > p$. This is observed by Conti via an example and is also observed in [A3] by noting that certain capacity externals belong to a Morrey Space, but because of the relations theorem for $C_{\alpha,p}$ capacities cannot belong to any higher order Lebesgue class, even locally.

Chapter 14
Morrey-Besov Spaces and Besov Capacity

14.1 Adams-Lewis inequality (Sobolev inequality for Morrey-Besov)

Here, we make a brief visit to the theory of Besov spaces, Besov capacity, and the Morrey-Besov analogue of Theorem 7.1 (i) - the Sobolev inequality for the Morrey-Besov setting. In fact, we shall say that a function $u(x)$ satisfies the Morrey-Besov condition if $u \in L^p(\mathbb{R}^n)$, $1 \leq p < \infty$ and

$$\left[\int_{\mathbb{R}^n} \left(\int_Q |\Delta_t^k u(x)|^p \, dx \right)^{q/p} |t|^{-(n+\alpha q)} \, dt \right]^{1/q} \leq A_0 \, |Q|^{(1-\lambda/n)/p}, \tag{14.1}$$

for each cube Q, $1 \leq q < \infty$, $\alpha > 0$. Here $\Delta_t^k u(x)$ is defined inductively, as the k-th order difference operator

$$\Delta_t^1 u(x) = u(x+t) - u(x)$$
$$\Delta_t^2 u(x) = \Delta_t^1 \Delta_t^1 u(x), \text{ etc.}$$

For the norm of (14.1). $k \in \mathbb{N}$ is chosen so that $k - 1 < \alpha \leq k$ and $0 < \lambda \leq n$. Note that when $\lambda = n$, (14.1) gives the standard Besov norm (quasi norm) for $u \in B_\alpha^{p,q}(\mathbb{R}^n)$.

© Springer International Publishing Switzerland 2015
D. Adams, *Morrey Spaces*, Applied and Numerical Harmonic Analysis,
DOI 10.1007/978-3-319-26681-7_14

Another interesting Besov type space is that of [X], a Holomorphic Q class on the boundary of the unit disc in $\mathbb{R}^2$. He requires

$$\sup_{I \subset \partial\Delta} |I|^{-p} \int_I \int_I \frac{|f(\zeta) - f(\eta)|^2}{|\zeta - \eta|^{2-p}} |d\zeta||d\eta| < \infty$$

$\partial\Delta$ = boundary of unit disc Δ. Notice that this space is smaller than $B^{2,2}_{\frac{1-p}{2}}(\mathbb{R}^1)$, $0 < p < 1$, when defined on $\partial\Delta$.

Our result here is

Theorem 14.1. Let p, q, k, α, λ be as above with $1 \le p < \lambda/\alpha$, $\tilde{p} = \lambda p/(\lambda - \alpha p)$, then any $u \in L^p(\mathbb{R}^n)$ that satisfies (14.1), also satisfies, for each cube Q

$$||u \cdot X_Q||_{L^{(\tilde{p}, q\tilde{p}/p)}} \le c \, A_0 \, |Q|^{(1 - \lambda/n)/\tilde{p}} \tag{14.2}$$

for some constant c independent of u, A_0, and Q.

In (14.2), the left side is being taken with respect to the Lorentz norm

$$||f||_{L^{(a,b)}} = \left(\int_0^\infty [s \, |\{x : |f(x)| > s\}|^{1/a}]^b \frac{ds}{s} \right)^{1/b},$$

$1 \le a, b < \infty$. Recall $L^{(a,a)} = L^a$, and $L^{(a,b_1)} \subset L^{(a,b_2)}$ when $b_1 \le b_2$.

Proof. First, notice that if $f(x,t) = |t|^{-\gamma} \Delta^k_t u(x)$, and $I^{(2n)}_\gamma$ = Riesz kernel on $\mathbb{R}^{2n}$, $\gamma = \alpha + n/q$, then via the Fourier transform of convolutions over $\mathbb{R}^{2n}$. we can write

$$u(x) = c \cdot I^{(2n)}_\gamma f(x, 0) \tag{14.3}$$

for a.e. $x \in \mathbb{R}^n$. And then it suffices to show

$$||X_Q \cdot I^{(2n)}_\gamma \, |f|(\cdot ; 0)||_{L^{\tilde{p}, q\tilde{p}/p}} \le c \, A_0 \, |Q|^{(1 - \lambda/n)\tilde{p}} \tag{14.4}$$

to achieve (14.2).

So to prove (14.4), write, for all $r > 0$,

$$V_x(|f|, r) = \iint_{|s|^2 + |y|^2 \le r^2} |f(y + x, s)| \, dy \, ds \le c \, A_0 \, r^{2n - \lambda/p - n/q} \tag{14.5}$$

via Hölder's inequality. And

$$I^{(2n)}_\gamma |f|(x, 0) = (2n - \gamma) \int_0^\infty V_x(|f|, r) r^{\gamma - 2n} \frac{dr}{r}. \tag{14.6}$$

Now let Q be a chosen cube, center at x_0 and edge length $|l(Q)| = \rho$, and set

$$F(x, t) = \begin{cases} |f(x,t)|, & |x - x_0|^2 + |t|^2 \le 4n\rho^2 \\ 0, & \text{otherwise.} \end{cases}$$

And $G(x,t) = |f(x,t)| - F(x,t)$. Then for $x \in Q$

$$I_\gamma^{(2n)} G(x,0) \le (2n - \gamma) \int_0^\infty V_x(|f|, r) \frac{dr}{r}$$

$$\le c\, A_0\, |Q|^{-\lambda/n\bar{p}} \tag{14.7}$$

follows from (14.5), $\gamma = \alpha + n/q$. And again from (14.5)

$$I_\gamma^{(2n)} F(x,0) = (2n - \gamma) \int_0^\infty r^{2n-\gamma} V_x(F, r) \frac{dr}{r}$$

$$= (2n - \gamma) \left[\int_0^\sigma \cdots + \int_\sigma^\infty \cdots \right]$$

$$\le c \left[\int_0^\sigma \cdots + A_0 \sigma^{-\lambda/\bar{p}} \right]. \tag{14.8}$$

Now set $E_s = \{x : I_\gamma^{(2n)} F(x,0) > s\}$ and $\sigma = (c_0\, s\, A_0^{-1})^{-\bar{p}/\lambda}$, then in (14.8)

$$s \cdot |E_s| \le \int_{E_s} I_\gamma^{(2n)} F(x,0)\, dx \le c \int_{E_s} \left(\int_0^\sigma \cdots \right) + c_0\, s|E_s|.$$

Thus for c_0 sufficiently small

$$s \cdot |E_s| \le c \int_0^\sigma r^{\gamma-2n} \int_{E_s} V_x(F, r)\, dx\, \frac{dr}{r}$$

$$\le c \int_0^\sigma r^{\gamma-2n} \left[\int_{|t|<r} \int_{|y|<r} \int_{E_s} F(x+y,t)\, dx\, dy\, dt \right] \frac{dr}{r}$$

$$\le c\, |E_s|^{1-1/p} \int_0^\sigma r^{\gamma-n} \left[\int_{|t|<r} \left(\int F(x,t)^p\, dx \right)^{1/p} dt \right] \frac{dr}{r}.$$

Hence if $g(t) = \left(\int F(x,t)^p\, dx \right)^{1/p}$, then

$$s\, |E_s|^{1/p} \le c \int_0^\sigma r^{\gamma-n} \left(\int_{|t|<r} g(t)\, dt \right) \frac{dr}{r}$$

$$\le \int_{|t|<\sigma} g(t) \int_{|t|}^\infty r^{\gamma-n+\epsilon} \frac{dr}{r}\, dt$$

$$\le \int_{|t|<\sigma} g(t)|t|^{-\epsilon}\, dt \cdot \sigma^{\alpha+n/q-n+\epsilon}, \tag{14.9}$$

provided $\epsilon > 0$ is chosen so that $n - n/q - \alpha < \epsilon < n - n/q$. Next, notice that

$$\sigma_s^{-\alpha} = (c_0 \, A_0^{-1})^{\alpha \tilde{p}/\lambda} \cdot s^{\tilde{p}/p}$$

and that $\dfrac{d\sigma}{\sigma} = -\dfrac{\tilde{p}}{\lambda} \cdot \dfrac{ds}{s}$.

Hence multiplying (14.9) by $(A_0 \, c_0^{-1})^{\alpha \tilde{p}/\lambda} \sigma^{-\alpha}$, raising to the power q and integrating yields

$$\int_0^\infty (s^{\tilde{p}} \, |E_s|)^{q/p} \frac{ds}{s} \leq c \, A_0^{\alpha \tilde{p}q/\lambda} \int_0^\infty \left(\sigma^{n/q-n+\epsilon} \int_{|t|<\sigma} g(t)|t|^{-s} \, dt \right)^q \frac{d\sigma}{\sigma}$$

$$\leq c \, A_0^{\alpha \tilde{p}q/\lambda} \int g(t)^q \, dt \leq c \, A_0^{(1+\alpha \tilde{p}q/\lambda)q} \, |Q|^{(1-\lambda/n)q/p}$$

$$\leq c \, A_0^{\tilde{p}q/p} \, |Q|^{(1-\lambda/n)q/p},$$

from Holder's inequality and the definition (14.1). Thus, we get

$$\|I_\gamma^{(2n)} F(\cdot \, , 0)\|_{L^{\tilde{p},q\tilde{p}/p}} \leq c \, A_0 \, |Q|^{(1-\lambda/n)/\tilde{p}}$$

and hence (14.4) follows.   □

Theorem 14.1 is taken from [AL] and was a response to an earlier paper by Ross [R], who did the case corresponding to $q = \infty$ - the so-called Morrey-Nikolski case. Also, the reader might note that in [AL] there are local versions of Theorem 14.1 where one estimates $(u-u_{Q_0})X_{Q_0}$, in the same Lorentz norm in terms of $|Q_0|^{(1-\lambda/n)/\tilde{p}}$

14.2 Besov capacity and the Netrusov capacity

Having come this far, it would be a shame not to mention some nice properties of Besov spaces and their capacities and associated Wolff potentials. For this it is best to refer the reader to [A9] and [AH]. But one nice connection should be mentioned, i.e., the relationship between the Besov capacities $C(\cdot \, , B_\alpha^{p,q})$ and the Netrusov capacities mentioned in Theorem (3.6) (a) $\Lambda^{d;\theta}(\cdot)$.

$$C(E; B_\alpha^{p,q}) = \inf\{\|u\|_{B_\alpha^{p,q}}^p : u \geq 1 \text{ on } E\}$$

Then

$$C(\cdot \, ; B_\alpha^{p,q}) \sim \Lambda^{n-\alpha p; q/p}(\cdot)$$

when either $p \leq 1$ or $q \leq 1$, $\alpha p < n$. In fact, it was Netrusov that finally found the relationship between the Sobolev capacities $\dot{C}_{\alpha,p}$ for $p \leq 1$. (now ones uses the

Hardy spaces to replace the L^p spaces when $p \leq 1$), and the Hausdorff capacities: $\dot{C}_{\alpha,p} \sim C(\cdot \; ; B_{\alpha}^{p,q}) \sim \Lambda^{n-\alpha p}(\cdot)$ for $p \leq 1$. But these issues are best left for another time. The reader is referred to [A9] and [AH].

14.3   Notes: CSI for Besov capacities

(1) Besov capacity and CSI: these results come from [AX1]. For $u \in \dot{B}_{\alpha}^{p,q}$ ((14.1) with $\lambda = n$):

(a)

$$\int_0^\infty C\left([|u| > t]; \dot{B}_{\alpha}^{p,q}\right)^{q/p} dt^q \leq c \, ||u||_{B_{\alpha}^{p,q}}^q \quad , 1 < p \leq q < \infty.$$

(b)

$$\int_0^\infty C\left([|u| > t]; \dot{B}_{\alpha}^{p,q}\right) dt^p \leq c \, ||u||_{B_{\alpha}^{p,q}}^p \quad , 1 < q \leq p < \infty.$$

Again these imply some trace results for u.

Chapter 15
Morrey Potentials and PDE I

In these next two chapters, we use much of our previous theory to examine the regularity of certain elliptic non-linear PDE in domains $\Omega \subset \mathbb{R}^n$. So in some sense, we have come full circle, back to the kind of PDEs that originally motivated Morrey in the first place to introduce his Morrey condition. In this chapter, we study the familiar equation: $-\Delta u = u^p$, for $p > n/(n-2)$, $u \geq 0$ on Ω; $\partial\Omega$ smooth. In Chapter 16, we will look at non-linear elliptic systems.

15.1 $-\Delta u = u^p$, $u \geq 0$

Consider now the problem of the regularity of weak solutions to

$$-\Delta u = u^p, \ u \geq 0 \text{ in } \Omega; \ p > 0. \tag{15.1}$$

See [Pa] and [SY]. u is a weak solution if $u \geq 0$ and for all $\varphi \in C_0^\infty(\Omega)$,

$$-\int u \, \Delta\varphi \, dx = \int u^p \, \varphi \, dx \tag{15.2}$$

where $u \in L^p(\Omega)$. Here Ω is a bounded domain with smooth boundary $\partial\Omega$.

The first results here remove $p < n/(n-2)$ from consideration since it is well known that the solution to (15.1) will automatically be $C_0^\infty(\Omega)$. We comment briefly on this below.

Theorem 15.1. For $1 \leq p < n/(n-2)$, $n \geq 3$, every non-negative solution to u to (15.2) is regular $\equiv C^\infty(\Omega)$.

© Springer International Publishing Switzerland 2015
D. Adams, *Morrey Spaces*, Applied and Numerical Harmonic Analysis,
DOI 10.1007/978-3-319-26681-7_15

For the remaining values for $p \geq 1$, we have

Theorem 15.2. Let u be a non-negative solution to (15.2) in Ω, then there is an open set $\Omega' \subset \Omega$ such that u is regular in Ω' and $\dot{C}_{2,p'}(\Omega \setminus \Omega') = 0$, $p > n/(n-2)$, $n \geq 3$.

Proof. The proofs of these two theorems take several steps to complete, and we first present the necessary tools for both. We have omitted the case $p = \frac{n}{n-2}$ for it needs special (limiting) considerations which we choose to leave for the reader.

Step 1

Theorem 15.3. If u is a non-negative solution to (15.2), then

$$u \in L^{p,2p'}(\Omega), \quad p' \leq n/2.$$

Proof. Let $\varphi(x) = \eta\left(\dfrac{x - x_0}{r}\right)^\sigma$ in (15.2), where $\eta \in C_0^\infty(B(0,1))^+$ and $\sigma > 2p$. Then

$$\int u^p \eta^\sigma \, dx \leq \frac{C}{r^2}\left(\int u^p \eta^\sigma \, dx\right)^{1/p} \cdot r^{n/p'},$$

hence

$$\int u^p \, \eta^\sigma \, dx \leq C \, r^{n-2p'}.$$

$\square$

Theorem 15.4. Suppose u is a non-negative solution to (15.2) in Ω, then there is a constant a_p such that for $x \in \Omega$, and $r > 0$, r is small enough;

$$\fint_{B(x,r)} u^p dy \leq a_p \left\{ \left(\fint_{B(x,2r)} u^{p-1} dy\right)^{p'} + \fint_{B(x,2r)} u(y)^p \left(\int_{B(y,2r)} |y-z|^{2-n} u(z)^{p-1} dz\right) dy\right\}$$

(15.3)

for $p \geq 2$, and

$$\fint_{B(x,r)} u^p dy \leq a_p \left\{ \left(\fint_{B(x,2r)} u dy\right)^{p} + \fint_{B(x,2r)} u(y)^p \left(\int_{B(y,2r)} |y-z|^{2-n} u(z)^{p-1} dz\right) dy\right\}$$

(15.4)

for $1 < p < 2$.

Proof. Let G be the Green function for Δ in the ball $B(x,r)$, then

$$-\Delta(u - Gu^p) = 0 \text{ in } \Omega.$$

Hence by the mean value property for Harmonic functions for $y \in \Omega$

$$u(y) - Gu^p(y) = \fint_{B(y,r)} u \, dz - \fint_{B(y,r)} Gu^p \, dz$$

or

$$u(y) \leq \fint_{B(y,r)} u \, dz + Gu^p(y)$$

or

$$u(y) \leq \fint_{B(y,r)} u \, dz + c \int_{|y-z|<r} |y-z|^{2-n} u(z)^p \, dz. \tag{15.5}$$

Now multiply through inequality (15.5) by u^{p-1} and then integrate over the ball $B(x,r)$, giving

$$\fint_{B(x,r)} u(y)^p \, dy \leq \int_{B(x,r)} u(y)^{p-1} \cdot \fint_{B(y,r)} u(z) \, dz \, dy$$

$$+ c \int_{B(x,r)} u(y)^{p-1} \int_{B(y,r)} |y-z|^{2-n} u(z)^p \, dz \, dy$$

$$\leq c \left(\int_{|x-z|<2r} u(z)^{p-1} \, dz \right)^{p'}$$

$$+ c \int_{|x-z|<2r} u(z)^p \int_{|y-z|<2r} |y-z|^{2-n} u(y)^{p-1} \, dy \, dz$$

for $p \geq 2$.

A similar argument gives the second statement of the theorem. $\square$

Step 2

Theorem 15.5. If u is a non-negative solution to (15.2) in Ω, and $u \in L^{p,\mu}(\Omega)$ for any $\mu < 2p'$, then $u \in C^\infty(\Omega)$.

Proof. Notice that u satisfies

$$u(x) = c \cdot I_2 u^p(x)$$

on $\mathbb{R}^n$ (extended by zero outside Ω). Then using the weak type version of 6.1 (i), we can write

$$u^p \in L^{1,\mu} \implies u \in L^{q_1,\mu}, \quad \mu > 2,$$

for $q_1 < \frac{\mu}{\mu-2}$. And

$$u^p \in L^{q_1/p,\mu} \implies u \in L^{q_2,\mu}, \ \mu > \frac{2}{p} + 2,$$

for $q_2 < \frac{\mu}{(\mu-2)p-2}$. And

$$u^p \in L^{q_2/p,\mu} \implies u \in L^{q_3,\mu}, \ \mu > 2 + \frac{2}{p} + \frac{2}{p^2},$$

for $q_3 < \frac{\mu}{(\mu-2)p^2-2p-2}$, etc.

After k steps,

$$u \in L^{q_k,\mu}, \ \mu > 2 + \frac{2}{p} + \cdots + \frac{2}{p^{k-1}},$$

for

$$q_k < \frac{\mu}{(\mu-2)p^k - 2\sum_{j=0}^{k-1} p^j}.$$

But since $2\sum_{j=0}^{\infty} \left(\frac{1}{p}\right)^j = 2p'$ and we are assuming that $\mu < 2p'$, eventually μ will be less than a finite sum meaning that u must be Hölder continuous (Morrey's Lemma). For a higher order regularity, we do the usual: differentiate the PDE and then apply the standard regularity techniques of elliptic PDE.   □

Next, note that $\int |x-y|^{2-n} u(y)^{p-1} \, dy$ is continuous for $p' > n/2$ and hence we can make

$$\int_{|x-y|<R} |x-y|^{2-n} u(y)^{p-1} \, dy < \epsilon$$

for any small $\epsilon > 0$, uniformly choosing R sufficiently small. When this is applied to (15.3) and (15.4), we see that the theory of Reverse Hölder estimates applies (see [BF], page 25) and yields:

there exists an exponent $q > p$ such that $u \in L^q_{loc}(\Omega)$.

This then implies

$$\int_{B(x,r)} u^p \, dy \leq c \left(\int_{B(x,r)} u^q \, dy \right)^{p/q} r^{n(1-p/q)}$$

so that

$$r^{2p'-n} \int_{B(x,r)} u^p \, dy \leq \|u\|_{L^q}^p \, r^{n-np/q}$$

and that $2p' - np/q > 0$ when $p' > \frac{n}{2}$; i.e., $p < \frac{n}{n-2}$.

This concludes the proof of Theorem 15.1.

Step 3

Next, we want to do something like the above except that $\int_{|x-y|<R} |x - y|^{2-n} u(y)^{p-1} \, dy$ may not be small for small R. So we need to estimate the set where

$$\zeta_u(x) \equiv \overline{\lim_{y \to x}} \int_{|y-z|<|x-y|} |y - z|^{2-n} u(z)^{p-1} \, dz$$

stays away from zero, i.e., we are interested in the set $\{x \in \Omega : \zeta_u(x) \geq \lambda\}$ for $\lambda > 0$. Also, notice that this set is the same as the set $\text{sing}(u) =$

$$\left\{ x \in \Omega : \overline{\lim_{y \to x}} \int_{|y-z|<|x-y|} |y - z|^{2-n} |u(z)^{p-1} - \varphi(z)| \, dz \geq \lambda \right\},$$

for any $\varphi \in C^\infty(\Omega)$. Furthermore,

$$\lambda \leq \zeta_u(x) \leq I_2(|u^{p-1} - \varphi|)(y) + \lambda/2$$

for y in some neighborhood $N(x)$, $x \in \text{sing}(u)$.
  Hence,

$$\dot{C}_{2,p'}\left(N(x) \cap \left\{ y \in \Omega : \int |y - z|^{2-n} |u(z)^{p-1} - \varphi(z)| dz \geq \lambda/2 \right\} \right)$$

$$\leq \left(\frac{2}{\lambda} \right)^{p'} \|u^{p-1} - \varphi\|_{L^{p'}}^{p'},$$

by the definition of $\dot{C}_{2,p'}$. Thus by taking φ to approximate u^{p-1} in $L^{p'}$, we arrive at the fact

$$\dot{C}_{2,p'}(\{x : \zeta_u(x) > 0\}) = 0.$$

So any neighborhood $B(x, r)$ in the compliment of $[\zeta_u \geq \lambda]$ must have $\zeta_u < \lambda$. This allows us to choose $\lambda = \frac{1}{2a_p}$, a_p the constant in Theorem 15.3.
  So, now we use a Lemma of [P] (similar to those from [G]).

Lemma 15.6. Suppose that

$$\int_{B(x,R)} |x - y|^{2-n} u(y)^{p-1} \, dy < \frac{1}{2a_p} \tag{15.6}$$

for all x in some neighborhood of $x_0 \in \Omega$, for R small enough, and that

$$\int_{B(x,R)} u(y)^p \, dy < \epsilon_0^p \, R^{n-2p'} \tag{15.7}$$

for ϵ_0 sufficiently small, then there is a constant $\Theta \in (0, 1)$ such that

$$\frac{1}{(\Theta R)^{n-2p'}} \int_{B(x,\Theta R)} u(y)^p \, dy \leq \frac{1}{2} \frac{1}{R^{n-2p'}} \int_{B(x,R)} u(y)^p \, dy. \qquad (15.8)$$

We will not prove this lemma here, but refer the reader to [P]. But notice that (15.7) is not really a restriction because Non-Linear Potential Theory gives:

$$I_2(I_2 \, u^p)^{p-1}(x) < \infty \implies \lim_{r \to 0} r^{2p'-n} \int_{B(x,r)} u(y)^p \, dy = 0,$$

and this holds $\dot{C}_{2,p'}$ - a.e. on Ω. In fact from [AH], it follows that

$$I_2(I_2 \, u^p)^{p-1}(x) \geq c \, W_{2,p'}^{u^p \, dy}(x)$$

the standard lower bound for non-linear potentials in terms of the Wolff potentials (see Chapter 6).

Now, using Lemma 15.6, we choose a point

$$x_0 \in \Omega \setminus s, \; s = \left\{ x \in \Omega : \int_{B(x,R)} u(y)^p \, dy \geq \epsilon_0^p \, R^{n-2p'}, \text{ all } R < R_0 \right\}$$

for some R_0, such that

$$\int_{B(x,R)} u(y)^p \, dy < \epsilon_0^p \, R^{n-2p'}, 0 < R < R_0$$

and for all x is some neighborhood of x_0. Then by Lemma 15.6, there exists $\mu < 2p'$ and $c > 0$ such that

$$\int_{B(x,R)} u(y)^p \, dy \leq c \, R^{n-\mu}$$

for all x in some neighborhood of x_0 and all $R < R_0$. This is determined by induction on (15.8), from which we can get

$$\frac{1}{(\theta^k R)^{n-2p'}} \int_{B(x,\theta^k R)} u(y)^p \, dy \leq 2^{-k} \, R^{2p'-n} \int_{B(x,R)} u(y)^p \, dy$$

for $k \in \mathbb{N}$. μ is now chosen such that

$$\theta^{\mu+2p'-n} > \frac{1}{2}.$$

$\square$

15.2 Notes

15.2.1 The Yamabe Case $p = \frac{n+2}{n-2}$

The case $p = \dfrac{n+2}{n-2}$ is usually referred in (15.1)–(15.2) as the Yamabe case. In fact, here the solution set becomes very rich, even for singular solutions; see [SY], where solutions to - $L_0\, u = u^{\frac{n+2}{n-2}}$, L_0 conformal Laplacian, are sought on domains $\Omega \subset S^n$. And [SY] demonstrates that there are singular sets of solutions of dimensions $= \frac{n-2}{2}$ (global singular sets), but one can also have singular sets of the Cantor type of small dimension. As we shall see, this behavior is not permitted with the systems of the next chapter, i.e., Harmonic map systems.

15.2.2 Stationary Navier-Stokes (n = 5)

It should also be mentioned in these notes the particular success of the Morrey theory for estimates of the Stationary Navier-Stokes equations due to Bensousson-Frehse [BF]. Their estimates are most notable in $n = 5$ dimensional space.

We shall say $u(x) \in W_0^{1,2}(\Omega)^n$, div $u = 0$, and $p(x) \in W^{1,n/(n-1)}(\Omega)$ satisfies the n-dimensional stationary Navier-Stokes system in

$$-\Delta u + u \cdot Du + Dv = f$$

$$-\Delta p = Du \cdot Dv - \operatorname{div} f$$

$$\omega = p + \frac{|u|^2}{2} \le c_0$$

for some constant c_0. Such a solution, when it exists, is termed regular if

$$u \in W_{loc}^{2,s}(\Omega)^n \text{ and } p \in W_{loc}^{1,s}(\Omega)$$

for all $s > 1$. [BF] term these maximum solutions when $n = 5$.

The interest to us in these notes is that a result like this is obtained by iterating the Morrey norm, initially shown to satisfy

$$|Du| \in L^{2,n-4+2\delta}$$

for some $\delta > 0$. It then follows that

$$u \in L^{2(\frac{2-\delta}{1-\delta}),\lambda} \text{ and } u \cdot Du \in L^{r_0,\lambda}$$

for $\lambda = n - 4 + 2\delta$ and $r_0 = 2(\frac{2-\delta}{3-2\delta})$ with similar estimates for the pressure p. Iterating (via Theorem 7.1) gives a remarkable result... that is best left to the experts, i.e., the discussion in Chapter 7 of [BF].

15.2.3   A Comment

Also, we should mention the work of M. Taylor [Ta] which uses the Morrey theory very effectively, again exploiting the Morrey-Sobolev inequality Theorem 7.1(i).

Chapter 16
Morrey Potentials and PDE II

16.1 Examples of singular sets for elliptic systems

First some examples of systems with isolated singular points; see [BF] and [G].
Let $u(x)$ be given by

$$u(x) = \left(u^1(x), u^2(x), \cdots, u^n(x) \right) = \frac{x}{|x|^\gamma},$$

where

$$x = (x_1, x_2, \cdots, x_n) \text{ and } \gamma > 0.$$

Then if we set

$$a_{ij}^{\alpha\beta}(x, u) = \delta_{ij}\delta_{\alpha\beta} + (c\delta_{i\alpha} + d \cdot b_{i\alpha}(x, u))(c\delta_{j\beta} + d \cdot b_{j\beta}(x, u))$$

c and d two positive constants, then in 1968 DeGiorgi showed that the above $u(x)$
solves

$$- (a_{ij}^{\alpha\beta}(x, u) u_{x_i}^\alpha)_{x_j} = 0, \quad \beta = 1, 2, \cdots, n \tag{16.1}$$

(summation convention) in the ball $B(0, 1)$ in the weak sense, when

$$b_{i\alpha}(x, u) = \frac{x_i x_\alpha}{|x|^2}, \tag{16.2}$$

with $\gamma = \frac{n}{2}(1 - 1/\sqrt{4(n-1)^2 + 1})$ and with $c = n - 2$, $d = n$. Then later,
Giusti-Miranda did the same, but now for

© Springer International Publishing Switzerland 2015
D. Adams, *Morrey Spaces*, Applied and Numerical Harmonic Analysis,
DOI 10.1007/978-3-319-26681-7_16

$$b_{i\alpha}(x, u) = \frac{u^i u^\alpha}{1 + |u|^2},\tag{16.3}$$

with $\gamma = 1$ and $c = 1$, $d = \frac{4}{n-2}$. In 1995, Koshelev refined the DeGiorgi example by showing that one can take

$$\gamma = 1, \quad c = (n-1)^{-1/2}\left(1 + \frac{(n-2)^2}{n-1}\right)^{-1/4}, \quad d = \frac{c + c^{-1}}{n-2},$$

in (16.1) with (16.2) in force. Of course, when $\gamma = 1$, our $u(x)$ is then a bounded discontinuous solution in $B(0, 1)$ with $x = 0$ the point of discontinuity. We next apply the Morrey theory we have developed to show that these examples (when $\gamma = 1$) represent a best possible scenario.

16.2   Meyers-Elcrat system

We now study the size of the exceptional (singular) set for the Meyers-Elcrat system, a 2m-th order quasi-linear elliptic system; see [BF]. We note that one is able to reduce the dimension of the singular set under the Sobolev regime [G], to $n - p$. Can one reduce it further under the Morrey-Sobolev regime?

One of the reasons that we concentrate on the Meyers-Elcrat system, is that every $W^{m,p}$ - solution u has a reverse-Hölder exponent $r > p$, such that

$$\left(\frac{\partial}{\partial x}\right)^m u \in L^r_{loc}(\Omega),$$

which then in turn implies that $u \in L^q_{loc}$ for some $q > np/(n - mp) =$ Sobolev exponent. For this, the reader is referred to [G] and [BF].

The Meyers-Elcrat system is:

$$\sum_{|\gamma| \le m} (-1)^{|\gamma|} \left(\frac{\partial}{\partial x}\right)^\gamma A_\gamma(x, D^m u) = 0, \text{ in } \Omega,\tag{16.4}$$

where

$$A_\gamma : \Omega \times \mathbb{R}^N \to \mathbb{R}^n, \quad N = \sum_{k=1}^m n^k,$$

and A_r is a Caratheodory function with

$$\begin{cases} \displaystyle\sum_{|\gamma|\le m} A_\gamma(x,D^m u)\cdot\left(\frac{\partial}{\partial x}\right)^\gamma u \ge a_0\left|\left(\frac{\partial}{\partial x}\right)^m u\right|^p, \text{ a. e. } \Omega, \\[4mm] \displaystyle |A_\gamma(x,D^m u)|\le M\cdot\left|\left(\frac{\partial}{\partial x}\right)^m u\right|^{p-1}, \quad |\gamma|\le m, \text{ a. e. } \Omega. \end{cases} \tag{16.5}$$

Here $1<p<\infty$, a_0 and M are positive constants and

$$D^m u = \left\{\left(\frac{\partial}{\partial x}\right)^\gamma u : |\gamma|\le M\right\}$$

$$\left(\frac{\partial}{\partial x}\right)^m u = \left\{\left(\frac{\partial}{\partial x}\right)^\gamma u : |\gamma| = m\right\}$$

$$\left(\frac{\partial}{\partial x}\right)^\gamma u = \left(\frac{\partial}{\partial x}\right)^{\gamma_1}\cdots\left(\frac{\partial}{\partial x}\right)^{\gamma_n}$$

$$\gamma = (\gamma_1,\cdots,\gamma_n)\in\mathbb{N}^n$$

$$|\gamma| = \gamma_1+\cdots+\gamma_n.$$

And for our main result, we define the singular set as

$$\sum_p(u,\Omega) = S_p(u,\Omega)\cup T(u,\Omega)$$

where

$$S_p(u,\Omega) = \left\{x\in\Omega : \overline{\lim_{r\to 0}}\fint_{B(x,r)}\left|u-\fint_{B(x,r)}u\right|^p dy > 0\right\}$$

and

$$T(u,\Omega) = \left\{x\in\Omega : \sup_{r>0}\left|\fint_{B(x,r)}u\,dy\right| = +\infty\right\}.$$

Theorem 16.1. Let Ω be a bounded domain in $\mathbb{R}^n$, then the compact singular set $\sum_p(u,\Omega)$ in Ω for $W^{m,p}$ solutions of (16.4), (16.5) are sets of $C_m(\cdot\,;L^{p,\mu})$ capacity zero, $d>n-mp$, $mp<\mu<n$.

Proof. It is enough to give the argument in the case $q<\infty$ for the other follows readily.

So, we begin by choosing the test function to be used in the usual weak or $W^{m,p}$ formulation of the solution of ((16.4)–(16.5)). We take $\varphi = \eta^{mp}\cdot u$, where $\eta(x) = \psi(\frac{x-x_0}{r})$, $x_0\in\Omega$, $\psi\in C_0^\infty(\mathbb{R}^n)$ with $\psi=1$ on $B(x_0,r/2)$ and zero outside $B(x_0,r)$. We then have, by the growth condition (16.5):

$$\int \left|\left(\frac{\partial}{\partial x}\right)^m u\right|^p \eta^{mp}\, dy \leq c \int \left|\left(\frac{\partial}{\partial x}\right)^m u\right|^{p-1} \cdot \eta^{m(p-1)} \left|\left(\frac{\partial}{\partial x}\right)^{m-k} u\right| r^{-(m-k)}\, dy$$

$$+ c \int \left|\left(\frac{\partial}{\partial x}\right)^m u\right|^{p-1} \cdot \eta^{m(p-1)} \left|\left(\frac{\partial}{\partial x}\right)^{j-l} u\right| r^{-(j-l)}\, dy$$

for $0 < k \leq m$, $j < m$, and $0 \leq l \leq j$. Then via Young's inequality $ab \leq \dfrac{a^p}{p\epsilon^p} + \dfrac{b^{p'}\epsilon^{p'}}{p'}$, we get

$$\int \left|\left(\frac{\partial}{\partial x}\right)^m u\right|^p \eta^{mp}\, dy \leq c \int_{B(x_0,r)} \left|\left(\frac{\partial}{\partial x}\right)^{m-k} u\right|^p \cdot r^{-(m-k)p}\, dy$$

$$+ \int_{B(x_0,r)} \left|\left(\frac{\partial}{\partial x}\right)^{j-l} u\right|^p r^{-(j-l)}\, dy.$$

We next use the Gagliardo-Nirenberg inequality (see [F]) to get

$$\int \left|\left(\frac{\partial}{\partial x}\right)^m u\right|^p \eta^{mp}\, dy \leq c\, r^{-mp} \int_{B(x_0,r)} |u|^p\, dy$$

$$\leq c\, r^{n-mp-np/q} \left(\int_{B(x_0,r)} |u|^q\, dy\right)^{p/q}.$$

This then gives

$$\left|\left(\frac{\partial}{\partial x}\right)^m u\right| \in L^{p,\lambda}, \quad \lambda = mp + \frac{np}{q}.$$

Next, to get an estimate on the singular set $\sum_p(u, \Omega)$, we first look at $f = |(\frac{\partial}{\partial x})^m u|$ and estimate the Morrey-Sobolev capacity of the set

$$E_t = \{x \in \Omega : [M_0(I_m f)^p]^{1/p}(x) > t\}$$

But notice

$$[M_0(I_m f)^p]^{1/p} \leq c\, I_m(M_0 f^p)^{1/p},$$

and then

$$C_m(E_t; L^{p,\mu}) \leq t^{-p}\|(M_0 f^p)^{1/p}\|^p_{L^{p,\mu}}$$

$$\leq c\, t^{-p}\|M_0 f^p\|_{L^{r,\theta}},$$

where $r = \frac{p+\epsilon}{p}$ and $\theta = \mu(\frac{p+\epsilon}{p})$. Thus

$$C_m(E_t; \ L^{p,\mu}) \leq c \, t^{-p} ||f^p||_{L^{r,\theta}},$$

since M_0 is a bounded operator on the Morrey spaces. And now we use the Reverse Hölder exponent $q > p$ setting $q = p + \epsilon$ and note

$$\left(\fint_{B(x_0,r)} f^q \, dy \right)^{1/q} \leq c \left(\fint_{B(x_0,r)} f^p \, dy \right)^{1/p}$$

to finally get

$$||f||_{L^{p+\epsilon,\theta}}^p \leq c \, ||f||_{L^{p,\mu}}^p.$$

Thus we have

$$C_m(E_t; \ L^{p,\mu}) \leq c \, t^{-p} ||f||_{L^{p,\mu}}^p. \tag{16.6}$$

We use (16.6) to estimate $C_m(S_p(u, \Omega); \ L^{p,\mu})$. Again, as in Chapter 15, the idea is to notice that

$$S_p(u, \Omega) = S_p(|u - \varphi|, \Omega)$$

for any smooth φ. But since $u \sim I_m \, (\frac{\partial}{\partial x})^m u$, it follows that

$$|u - \varphi| \leq c \left| I_m \left(\frac{\partial}{\partial x} \right)^m u - I_m \psi \right|$$

upon writing $\varphi = I_m \psi$. But then

$$\varlimsup_{r \to 0} \fint_{B(x,r)} \left| (u - \varphi) - \fint_{B(x,r)} (u - \varphi) \, dy \right|^p \, dy \leq c \, M_0(|u - \varphi|^p).$$

And so

$$C_m([M_0|u - \varphi|^p > t]; \ L^{p,\mu}) \leq c \, t^{-p} \left\| \left(\frac{\partial}{\partial x} \right)^m u - \psi \right\|_{L^{p,\mu}}^p \tag{16.7}$$

by the above. By now taking $\mu > \lambda$, Theorems 2.3 and 2.4 imply that the norm in (16.7) can be made arbitrarily small. Hence $C_m(S_p(u, \Omega); \ L^{p,\mu}) = 0$ for all $\mu > \lambda$. $T(u, \Omega)$ can be treated similarly. $\square$

16.3 Notes

16.3.1 Harmonic Maps

In fact this argument can be used to decide the nature of local singularities for
Harmonic maps. Here, a Harmonic map (see [BF, LW]) for our purposes, is a
minimizer of the functional

$$T(u, \Omega) = \int_{\Omega} A_{ij}^{\alpha p}(x, u) \, u_{x_\alpha}^i \, u_{x_\beta}^j \, dx,$$

(summation convention) with $u = (u^1, u^2, \cdots, u^n)$ and

symmetry: $A_{ij}^{\alpha\beta} = A_{ji}^{\beta\alpha}$

boundedness: $|A_{ij}^{\alpha\beta}| \leq M = $ constant independent of x and u,

ellipticity: $A_{ij}^{\alpha\beta} \xi_\alpha^i \xi_\beta^j \geq a_0 |\xi|^2$,

Hölder coefficients : $|A_{ij}^{\alpha\beta}(x, z) - A_{ij}^{\alpha\beta}(x', z')| \leq c|(x - x', \, z - z')|^\theta$ for some
$\theta \in (0, 1)$,

splitting coefficients: $A_{ij}^{\alpha\beta}(x, u) = g_{ij}(x, u) \cdot G^{\alpha\beta}(x)$.

Such a system has a well-known property called the "monotone inequality," i.e.,
$B(x_0, s) \subset B(x_0, r)$ gives

$$r^{2-n} \int_{B(x_0, r)} |\nabla u|^2 \, dx - s^{2-n} \int_{B(x_0, s)} |\nabla u|^2 \, dx \geq 0.$$

But such an inequality immediately implies that

$$|\nabla u| \in L^{2,2}(\Omega),$$

since $u \in W^{1,2} \cap L^\infty$ are the usual assumptions on the possible minimizers. Note
that such maps can have global singularities- that extend to the boundary of Ω. Such
singularities can have dimension at most $n - 3$. An example is

$$\frac{\bar{x}}{|\bar{x}|}, \; x = (\bar{x}, \hat{x}), \; \bar{x} \in \mathbb{R}^3, \; \hat{x} \in \mathbb{R}^{n-3}.$$

See [LW] and [Si]. In fact, Leon Simon gives an excellent accounting of such
manifold-singularities in the general case. Another description of these singularities
is given in [AX5] in the case when the range of the Harmonic map is a sphere S^{n-1}:
they can be realized as infinities of the Riesz potential of the Hilbert-Schmidt norm
of $\frac{\partial}{\partial x} u$, namely

$$\left[\sum_{i,j=1}^{n} \left| \frac{\partial}{\partial x_j} u^i \right|^2 \right]^{1/2}.$$

16.3.2 Lane-Emden systems

Further applications to PDE are given in [AX6], in particular to the Lane-Emden system of (p, q) type:

$$-\mathrm{div}(|\nabla u|^{p-2}\nabla u) \ = \ |u|^{q-1}u$$

where $u \in W^{1,p}(\Omega; \mathbb{R}^n) \cap L^q(\Omega; \mathbb{R}^n)$. And again the size of $\sum_L(u, \Omega)$ is estimated. We leave the interested reader to pursue this example as an application of the Morrey theory developed in these chapters and challenge the reader to further apply our theory to such PDE.

Chapter 17
Morrey Spaces On Complete Riemannian Manifolds

Because of the recent activity in Riemannian Geometry: solving nonlinear PDE that arise - specifically say the classic Gaussian curvature equation when prescribing a new metric via conformal change and similar questions in higher dimensions (see [AC]), or determining what properties of a non-compact manifold are needed to admit a Sobolev inequality/Sobolev embedding on that manifold, it seems like a good idea to extend our Morrey Theory to this setting. Specifically, one wants this refined Sobolev theory to again deal with the nature of singular sets, as in Chapter 16.

The history of Sobolev inequalities on CRM (Complete Riemannian Manifolds) goes back to at least N. Varopolous and the Paris school of the 1980s, and most recently to the work of Saloff-Coste and Hebey; see their recent lecture notes: [SC, H1], and [H2]. Here we find considerable progress not only in determining when a manifold (M^n, g) supports a Sobolev inequality, but also in determining the best constants in these inequalities/embeddings. We will not try to describe this theory, but mainly to note when we can extend our theory to (non-compact) CRMs.

Our first concern is to extend the Morrey-Sobolev inequality (6.1(i)). Typical assumptions on the manifold include

(i) (M^n, g) is complete - i.e., "geodesically complete"; one can connect any two points on the manifold by a minimal geodesic (Hopf-Rinow Theorem);
(ii) (M^n, g) has maximal volume growth; i.e.,

$$\mathrm{Vol}_g(B(x, r)) \geq c\, r^n, \quad \text{all } r > 0.$$

a similar upper bound holds when Ricci curvature is non-negative.
(iii) The Ricci curvature of M^n is non-negative, or possibly bounded below (for embeddings).

© Springer International Publishing Switzerland 2015
D. Adams, *Morrey Spaces*, Applied and Numerical Harmonic Analysis,
DOI 10.1007/978-3-319-26681-7_17

Our main conjecture here is that these three conditions are also sufficient for a Morrey-Sobolev inequality on (M^n, g). We will give supporting evidence below for this. Here the reader should be informed that we will be following closely the lead of [SC].

17.1   A counterexample

First of all, it is immediate that for an embedding $W^{1,p}(M) \subset L^q(M)$ one needs a constant $\theta > 0$ such that

$$\text{Vol}_g(B(x, r)) \geq \theta = \theta(M, p, r), \tag{17.1}$$

for all $x \in M$ and $r > 0$; $\frac{1}{q} = \frac{1}{p} - \frac{1}{n}$. See [H2], Lemma 2.2. So it is natural that if we construct a manifold for which (17.1) fails, no Sobolev embedding can hold. Such a manifold is

$$M = \mathbb{R} \times S^{n-1}, \quad g(x, \xi) = \mathcal{J}_x + u(x)h_\xi$$

where $\mathcal{J}_x$ is the Euclidean metric on $\mathbb{R}$, h is the standard metric on the unit sphere S^{n-1} of $\mathbb{R}^n$, and $u : \mathbb{R} \to (0, 1]$ is smooth and such that $u = 1$, when $t \leq 0$, $u(t) = e^{-2t}$ when $t \geq 1$. This manifold is complete, but fails to satisfy (17.1). Note that the Ricci curvature in this case is bounded below

$$\text{Ric}_{(M,g)} \geq -(n-1)g, \text{ for } t \geq 1.$$

see [H2], proposition 3.4.

Thus condition (ii) is necessary, but perhaps not condition (iii).

17.2   A Morrey-Sobolev inequality on M^n with balls having maximal growth and Ric ≥ 0

So we are now committed to extending our theory to find an analogue of our Morrey-Sobolev inequality in (M^n, g): assume $u \in L_{(M)}^{\tilde{p}, \lambda}$, $\tilde{p} = \lambda p / (\lambda - p)$, $1 < p < \lambda$?

Morrey Spaces are clearly well defined on a CRM: let $V(x, r) = \text{Vol}_g(B(x, r))$, and set $||f||_{L_{(M)}^{p, \lambda}}$ to be the $1/p$-th root

$$\sup_{x \in M^n, r > 0} \frac{1}{V(x.r)^{1 - \lambda/n}} \int_{B(x,r)} |f(y)|^p \, dy < \infty \tag{17.2}$$

where $B(x, r)$ is a ball $\subset M^n$ under the metric g. Of course, $0 < \lambda \leq n$, $1 < p < \infty$, and dy is standard volume measure on the (smooth) M^n.

The first order of business would then seem to be to extend the duality theory to M^n; i.e., $H^{p,\lambda}(M)$. There should be no difficulties here for the maximal operator extends to M^n for the volume measure is a "doubling measure." So again I would expect

$$\left(H^{p',\lambda}(M)\right)^* = L^{p,\lambda}(M),$$

but in this chapter, these are points left open - for the reader to think about! Also, expected is

$$X^{p,\lambda}(M) = K^{p,\lambda}(M) = Z^{p,\lambda}(M),$$

and the atomic decomposition for $L^1(\Lambda^d; M)$. With these, there is evidence that a Morrey-Sobolev inequality holds on M^n satisfying properties (i), (ii), (iii) above.

Following Stein [St2], we need to establish the duality inequality on (M^n, g):

$$\left|\int_M f(x)g(x)\,dx\right| \le c \int_M f^{\#}(x) \cdot M_0 g(x)\,dx \qquad (17.3)$$

for $f \in C_0^\infty(M)$ and $g \in H^{p',\lambda}(M)$. However, the main difficulty in achieving our goal is the representation of a smooth function in terms of its gradient over the entire manifold. This is the reason we need (17.3), so that we can proceed locally.

And following [SC], we can write

$$\left|V(x,r)f(x) - \int_{B(x,r)} f\,dy\right| \le \int_{B(x,r)} |f(x) - f(y)|\,dy \qquad (17.4)$$

$$\le \frac{r^n}{n} \int_{B(x,r)} |\nabla f(y)|\,d(x,y)^{1-n}\,dy,$$

where $d(x,y)$ is the distance on M^n between x and y. The left side of (17.4) is just

$$V(x,r)\left|f(x) - \frac{1}{V(x,r)}\int_{B(x,r)} f(y)\,dy\right|$$

and the maximal volume growth gives

$$f^{\#}(x) \le c \int_{B(x,r)} |\nabla f(y)|\,d(x,y)^{1-n}\,dy. \qquad (17.5)$$

At this point, we again estimate (17.5) using the maximal function technique we employed in Chapter 7, namely the RHS of (17.5) does not exceed

$$c\||\nabla f|\|_{L^{p,\lambda}(M)}^{p/\lambda} \cdot M_0(|\nabla f|)^{1-p/\lambda}, \quad \text{a.e.}$$

which just leads to

$$\|f^{\#}\|_{L^{\tilde{p},\lambda}} \leq c \, \| \, |\nabla f| \, \|_{L^{p,\lambda}},$$

$\tilde{p} = \lambda p/(\lambda - p)$, $1 < p < \lambda \leq n$. So it comes down to showing:
if $f^{\#} \in L^{p,\lambda}(M)$ and $f^{\#} \in L^{\tilde{p},\lambda}(M)$, then

$$\|f\|_{L^{\tilde{p},\lambda}(M)} \leq c\|f^{\#}\|_{L^{\tilde{p},\lambda}}.$$

But now we use duality

$$\left| \int_{m} f(x)g(x) \, dx \right| \leq c \int_{M} f^{\#}(x) \cdot M_0 g(x) \, dx$$

$$\leq c \, \|f^{\#}\|_{L^{\tilde{p},\lambda}(M)} \, \|M_0 g\|_{H^{\tilde{p}',\lambda}(M)}$$

$$\leq c \, \| \, |\nabla f| \, \|_{L^{p,\lambda}(M)} \cdot \|g\|_{H^{\tilde{p},\lambda}(M)}$$

giving our desired

$$\|f\|_{L^{\tilde{p},\lambda}(M)} \leq c \, \| \, |\nabla f| \, \|_{L^{p,\lambda}(M)}.$$

17.3   Further embedding and speculations

There are many details to be treated here, but this seems to be a way forward for Morrey-Sobolev theory on a Complete Riemannian Manifold - with maximal volume growth and non-negative Ricci curvature.

Other speculations here might involve the extension of this Morrey theory to the even more general setting of metric-measure spaces, especially in light of the recent potential theory extension given in[BB].

Further considerations:

(a) Use the duality of Theorem 5.6 to prove various results here on Manifolds connected with the space $\mathscr{L}^{p,\lambda}$.
(b) Use the ideas of Garnett-Jones to find the "distance in $\mathscr{L}^{p,\lambda}$ to $Ł^{p,\lambda}$." This is the Garnett-Jones problem when $\lambda = 0$, but this distance is zero for $\lambda > 0$. What is perhaps interesting is a question about approximating functions in $\mathscr{L}^{p,\lambda}$. Note if $f \in \mathscr{L}^{p,\lambda}$

$$(*) \quad \|I_{\lambda/p}f - \varphi\|_{\mathrm{BMO}} \leq c \, \|f - I_{-\lambda/p}\varphi\|_{\mathscr{L}^{p,\lambda}}, \quad \varphi \in L^{\infty}.$$

So it appears that approximating f by the distribution $I_{-\lambda/p}\varphi$ is not possible for $\varphi \in L_0^{\infty}$. Note if $\lambda/p = 2$, then $I_{-\lambda/p}\varphi = (-\Delta)\varphi$ and the (*) LHS cannot be made small by the G-J Theorem.

Bibliography

1. Papers by the author:

[A1] Traces of potentials arising from translation invariant operators, Annali Scuola Norm. Sup. Pisa 25(1971), 203–217.

[A2] A trace inequality for generalized potentials, Studia Math. 48(1973), 99–105.

[A3] Traces of Potentials II, Ind, U. Math. J. 22(1973), 907–918.

[A4] A note on Riesz potentials, Duke Math. J. 42(1975), 765–778.

[A5] The existence of capacitary strong-type estimates in $\mathbb{R}^n$, Ark. Math. 14(1976), 125–140.

[A6] Lectures on L^p-potential theory, Umeå Univer. Reports, August 1981.

[A7] A note on Choquet integrals with respect to Hausdorff capacity, Function Spaces and Appl., Proc. Lund 1986, Lecture Notes in Math., Springer 1988.

[A8] Weighted capacity and the Choquet integral, Proc. Amer. Math. Soc. 1023 (1998), 879–887.

[A9] Choquet integrals in potential theory, Pub. Mat. 42(1998), 3–66.

[A10] Besov capacity redux, Problems in Math. Analy. Vol. 42 (Russian); J. Math. Sci., Springer Vol 162 (2009) (English).

[A11] On F. Pacard's regularity for $-\Delta u = u^p$, EJDE 125(2012).

[A12] Mock Morrey Spaces, Proc. AMS 2013.

[A13] Dimension estimates for singular sets for focusing Yambe- type equations, Comm. PDE, 2015.

[A14] Estimates for the Hausdorff Dimension of the Blow up Sets for Semi-Linear Wave Equations, in preparation.

[A15] Bochner- Riesz Means of Morrey Functions, submitted to Proc. AMS, to appear.

2. Recent papers by the author and J. Xiao.

[AX1] Strong type estimates for homogeneous Besov Capacities, Math. Ann. 325(2003), 695–697.

[AX2] Nonlinear potential analysis on Morrey spaces and their capacities, Ind. U. Math. J. 53(2004), 1631–1666.

[AX3] Morrey Spaces in Harmonic Analysis, Arkiv Mat. 50(2012), 201–230.

© Springer International Publishing Switzerland 2015
D. Adams, *Morrey Spaces*, Applied and Numerical Harmonic Analysis,
DOI 10.1007/978-3-319-26681-7

[AX4] Morrey Potentials and Harmonic Maps, Comm. Math. Phys. 308(2011), 439–456.
[AX5] Regularity of Morrey Commutators, Trans. AMS 364(2012), 4801–4818.
[AX6] Singularities of Non-linear Elliptic Systems, Comm. PDE 38(2013), 1256–1273.
[AX7] Restrictions of Riesz-Morrey Potentials, in preparation.

3. Papers by the author and others.

[AM] Adams, D.R., Meyers, N. G. Thinness and Wiener criteria for nonlinear potentials, Ind. U. Math. J. 221 (1972), 169–197.
[AP] _____, Pierre, M., Capacitary strong type estimates in semilinear problems, Ann. Insti. Fourier, 41(1991), 117–135.
[AH] _____, Hedberg, L. I., Function spaces and potential theory, Gundlehren. #314, Springer, 1996.
[AL] _____, Lewis, J. L., On Morrey-Besov inequalities, Studia Math. 74(1982), 169–182.
[AE] _____, Eiderman, V., Singular operators, antisymmetric kernels, related capacities and Wolff potentials, Inter. Math. Res. Notes, 2013.

4. Further Bibliography

[An] Anger, B., Representation of capacities, Math. Ann. 229(1997), 245–258.
[BB] Bjorn, A., Bjorn, J., Nonlinear potential theory on metric spaces, Tracts in Math 17, European Math. Soc. 2011.
[BF] Bensoussan, A. , Frehse, J., Regularity results for nonlinear elliptic systems and appl., Springer 2002.
[BS] Bennett, C., Sharpley, R., Interpolation of operators, Pure Appl. Math. 129, Academic press 1988.
[BRV] Blasco, O., Ruiz, A., Vega, L., Non interpolation in Morrey-Campanato and block spaces, Ann. Scuola Norm. Sup. Pisa, 28(1999), 31–40.
[CF] Chiarenza, F., Frasca, M., Morrey spaces and Hardy-Littlewood maximal function, Rend. Mat. Appl. 3–4 (1998), 273–279.
[Ca] Campanato, S., Propertia di inclusione per spazi di Morrey, Ricerche Mat. 12(1963), 67–80.
[Co] Conti, F., Su alcuni spazi funzionali e loro applicazioni ad equazioni differenziali di tipo elliptico, Bull. Un. Mat. Ital. 4(1969), 554–569.
[C] Choquet, G., Theory of capacities, Ann. Inst. Fourier Grenoble, 5(1953), 131–295.
[Cl] Carleson, L., Selected problems on exceptional sets, Van Nostrand 1967.
[Ch] Chanillo, S., A note on commutators, Ind. U. Math. J., 31(1982), 7–16.
[CDW] Chen, Y., Ding, Y., Wang, X., Compactness of commutators of Riesz potential on Morrey spaces, Potential Ana., 30(2009), 301–313.
[CRW] Coifman, R., Rochberg, R. Weiss, G., Factorization theorems for Hardy-Spaces in several variables, Ann. Math. 103(1976), 611–635.
[D] Ding, Y., A characterization of BMO via commutators for some operators, Northeast. Math. J. 13(1997), 422–424.
[Da] Dahlberg, B., Regularity properties of Riesz potentials, Ind. U. Math. J. 28(1979), 257–268.
[F] Friedman, A., Partial differential equations, Holt- Rinehart-Winton 1961.

[FR] Fazio, di-Ragusa, M., Commutators and Morrey spaces, Bul. U. Math. Ital. 5(1991), 323–332.

[GF] Garcia-Cuerva, J. , de Franco, R., Weighted norm inequalities and related topics, 116 North Holland Math. Studies, North Holland 1985.

[G] Giaquinta, M., Multiple integrals in the calculus of variations and nonlinear elliptic systems, Ann. Math. Studies 105, Princeton Univ. Press. 1983.

[GJ] Garnett, J. , Jones, P., The distance in BMO to L^∞, Ann. Math., 108(1978), 373–393.

[H] Hansson, K., Imbedding theorems of Sobolev type in potential theory, Mathematica Scand. 45(1979), 77–102.

[He] Hedberg, L. I., On certain convolution inequalities, Proc. AMS. 36(1972), 505–510.

[HKM] Heinonen, J., Kilpelainen, T., Martio, O., Nonlinear Potential Theory of Degenerate Elliptic Equations, Oxford University Press, Oxford 1993.

[K] Kalita, E., Dual Morrey spaces, Dokl. Akad Nauk 361(1998), 447–449.

[KM] Komori, Y., Mizuhara, T. , Notes on commutators and Morrey spaces, Hokkaido Math. J. 32(2003), 345–353.

[LW] Lin, F., Wang, C., The analysis of harmonic maps and their heat flow, World Scientific 2008.

[Ma1] Maz'ya, V., A theorem on the multi-dimensional Schrodinger operator, Izv. Akad. Nauk. 28(1964), 1145–1172.

[Ma2] _____, On capacitory strong type estimates for fractional norms, Zap. Nauchn, Sem. LOMI 73(1977), 161–168.

[Mo] Morrey, C., On the solutions of quasi-linear elliptic partial differential equations, Trans. AMS 43(1938).

[M1] Meyers, N. G., A L^p - estimate for the gradient of solutions of second order elliptic divergence equations, Ann. Sc. Norm. Pisa 17(1963), 189–206.

[M2] _____, A theory of capacities for potentials of functions in Lebesgue classes, Math. Scand. 26(1970).

[ME] _____,Elcrat, A., Some results on regularity for solutions of nonlinear elliptic systems and quasi-regular functions, Duke Math. J. 42(1975), 121–156.

[MS1] Maźya, V. G., Shaposhnikova, T., Theory of multipliers in spaces of differentiable functions, Pitman Press 1985.

[MS2] _____, _____, Theory of Sobolev multipliers: with applications to differential and integral operators, 337 Grundlehren, Springer 2009.

[N1] Netrusov, Y., Metric estimates of capacities of sets in Besov spaces, Proc. Steklov Inst. (1992), 167–192.

[N2] _____, Estimates of capacities associated with Besov spaces, J. Soviet Math. 1996.

[Ni] Nieminen, E., Hausdorff measures, capacities, and Sobolev spaces with weights, Ann. Acad. Sci. Finland, Math. Dissertations, Helsinki 1991.

[OV] Orobitg, J. Verdera, J., Choquet integrals, Hausdorff content and the Hardy-Littlewood maximal operator, Bull. London Math. Society, 30(1998), 145–150.

[Pe1] Peetre, J., On the theory of $Ł_{p,\lambda}$ spaces, J. Funct. Anal. 4(1969), 71–87.

[Pe2] _____, On convolution operators leaving $L^{p,\lambda}$ spaces invariant, Ann. Math. Pura Appl. 72(1966), 295–304.

[Pa] Pacard, F., A regularity criterion for positive weak solutions of $-\Delta u = u^\alpha$, Comment. Math. Helvetici 68(1993), 73–84.

[R] Ross, J., A Morrey-Nikolski inequality, Proc AMS 8(1980), 97–102.

[ST] Sawano, Y., Tanaka, H., Predual spaces of Morrey spaces with non-doubling measures, Tokyo J. Math. 32(2009), 471–486.

[St1] Stein, E. M., Singular integrals and differentiability properties of functions, Princeton university press, 1970.

[St2] _____, Harmonic Analysis: real-variable methods, orthogonality, and oscillatory integrals, Princeton U. Press 1993.

[SZ] _____, Zygmund, A., Boundedness of translation invariant operators on Holder spaces and L^p-spaces, Ann. Math. 85(1967), 337–349.

[S] Stampacchia, G., The spaces $L^{p,\lambda}, N^{p,\lambda}$ and interpolation, Ann. Sco. Norm. Sup. Pisa 19(1965), 443–462.

[SY] Schoen, R., Yau, S. T., Conformally flat manifolds, Kleinian groups and scalar curvature, Invent. Math. 92(1988), 47–71.

[Si] Simon, L., Singularities of geometric variational problems, Nonlinear partial differential equations in differential geometry, Park City Math. Series, 2 AMS 1996.

[To] Torchinsky, A., Real-variable methods in harmonic analysis, 123 Pure & Appl. Math. series, Academic Press 1986.

[Tr] Troianiello, G., Elliptic differential equations and obstacle problems, Univ. series Math., Plenum Press 1986.

[Tu] Turesson, B. O., Nonlinear Potential Theory and Weighted Sobolev Spaces, Lecture Notes in Mathematics 1736, Springer 2000.

[U] Uchiyama, A., On the compactness of operators of Hankel type, Tohoku Math. J., 30(1978), 163–171.

[W] Wallin, H., Existence and Properties of Riesz Potentials Satisfying Lipschitz Conditions, Math. Scand. 19(1966), 151–160.

[Y^2] Yang, D., Yuan, W., A note on dyadic Hausdorff capacities, Bull. sci. math., 132(2008), 500–509.

[Y^{2}Z] _____, Zhuo, C., Complex interpolation on Besov-type and Triebel-Lizorkin-type spaces, Annal Appl. (Singap) , 11(2013).

[Z] Zorko, C., Morrey spaces, Proc. AMS 98(1986), 586–592.

[Ta] Taylor, M., Analysis on Morrey spaces and applications to Navier-Stokes and other evolution equations, Comm. PDE, 17(1992), 1407–1456.

[AC] Chang, A., Non-linear elliptic equations in conformal geometry, Zurich Lectures in Advanced Math., European Math. Soc., 2004.

[LY^2] Lu, Y., Yang, D., Yuan, W., Interpolation of Morrey Spaces on Metric Measure Spaces.

[X] Xiao, J., Holomorphic Q classes, Lecture Notes in Math., Springer #1767, 2001.

5. Generalized Morrey Spaces and Applications

[BG] Burenkov, V., Guliyev, V., Necessary and sufficient conditions for the boundedness of the Riesz potential in local Morrey-type spaces, to appear 2012.

[DK] Dzhabrailov, M., Khaligova, S., Anisotropic Fractional Maximal Operator in Anisotropic Generalized Morrey Spaces, J. Math. Research 4(2012).

[Gu] Guliyev, V., Boundedness of the maximal, potential and singular operators in the generalized Morrey spaces, J. Ineq. Appl. 2009.

[Maz] Mazzucato, A., Besov-Morrey spaces: function space theory and applications to nonlinear PDE, Trans. AMS. 355(2003), 1297–1364.

[Sh] Shen, Z., The periodic Schrodinger operators with potentials in the Morrey class, J. Funct. Anal., 193(2002), 314–345.

[GuM] Guliyev, V., Mammadov, Y., Riesz potential on the Heisenberg group and modified Morrey spaces, Analele Universitatii" Ovidius" Constanta-Seria Matematica 20(2012), 189–212.

[GM] Gogatishvili, A., Mustafayev, R., Equivalence of norms of Riesz potential and fractional maximal function in generalized Morrey spaces, Collect. Math. 63(2012), 11–28.

[LSY] Liu, L. Sawano, Y., Yang, D., Morrey-type spaces on Gauss measure spaces and boundedness of singular integrals, J. Geometric Anal. 2013, 1–45.

[LR] Lemarié-Rieusset, Potential Analysis, 38(2013), 741–752.

6. Riemannian Manifolds and Ricci Curvature

[SC] Saloff-Coste, L., Aspects of Sobolev-type inequalities, 289 Cambridge University Press 2002.

[H1] Hebey, E., Sobolev spaces on Riemannian manifolds, Lecture Notes Math. 1635, 1996, Springer.

[H2] _____, Nonlinear analysis on manifolds: Sobolev spaces and inequalities, Courant Lecture Notes, AMS 2000.

[GHL] Gallot, S., Hulin, D., Lafontaine, J., Riemannian geometry, third ed. Univ. Text., Springer, 2004.

[P] Perelman, G., Manifolds of positive Ricci curvature with almost maximal volume, J. AMS 7(1994), 299–305.

[CC] Cheeger, J., Colding, T., On the structure of spaces with Ricci curvature bounded below. I, J. Diff. Geom. 45(1997), 406–480.

Index of Symbols

© Springer International Publishing Switzerland 2015
D. Adams, *Morrey Spaces*, Applied and Numerical Harmonic Analysis,
DOI 10.1007/978-3-319-26681-7

121

Index

© Springer International Publishing Switzerland 2015
D. Adams, *Morrey Spaces*, Applied and Numerical Harmonic Analysis,
DOI 10.1007/978-3-319-26681-7